The Agriculture of China

Centre for Agricultural Strategy Series

Editors: C. R. W. Spedding and L. J. Peel

*Published in conjunction with the Centre
for Agricultural Strategy, University of Reading*

THE AGRICULTURE OF
China

Edited by
XU GUOHUA
Beijing Agricultural Engineering University

and

L. J. PEEL
University of Reading

OXFORD UNIVERSITY PRESS
1991

Oxford University Press, Walton Street, Oxford OX2 6DP

Oxford New York Toronto
Delhi Bombay Calcutta Madras Karachi
Petaling Jaya Singapore Hong Kong Tokyo
Nairobi Dar es Salaam Cape Town
Melbourne Auckland

and associated companies in
Berlin Ibadan

Oxford is a trade mark of Oxford University Press

Published in the United States
by Oxford University Press, New York

A catalogue record for this book is available from the British Library

ISBN 0–19–859208–6

Library of Congress Cataloging in Publication Data
Data available

ISBN 0–19–859208–6

Typeset by Colset Private Limited, Singapore

Printed and bound in
Great Britain by Biddles Ltd,
Guildford and King's Lynn

In memory of my dear wife Xue Ximeng who suffered a hard life for twenty years because of my unpleasant experiences, and died suddenly while our situation was improving.

Xu Guohua

Foreword

This is one of a series of volumes on the agriculture of different countries.

It is a generalist's book, written to provide both background information and a wider understanding of the agriculture of China in all its facets. It does not follow any particular disciplinary approach, but considers the farmer and his family, the systems of production he employs, the physical and environmental constraints on his activities, the economic system within which he must operate, and the historical and cultural backgrounds which help to determine all these things.

The book is intended to be useful to all those who wish to go to China for professional or vocational reasons connected with agricultural development, or trade in agricultural and related products; we hope that it may also assist those who, as tourists, want to learn more about local circumstances, and enlighten those who simply wish, without visiting China, to discover more about Chinese agriculture and rural life. This is why we are particularly concerned to provide the sort of general picture and breadth of understanding to which individuals can relate their own specific activities and interests. For agriculture as experienced by the farmer has many aspects, and those who would help him or trade with him must first understand something of the world as he sees it.

C.R.W.S
L.J.P.

Preface

Until almost the end of the 1980s a growing number of expatriates were travelling to China to work or advise on agricultural development projects. Increasing numbers of businessmen with an interest in agriculture, and of farmers and other tourists on special agricultural tours, were also visiting the country. This book has been designed to provide a source of background information on Chinese agriculture for similar people who may be travelling to China in the future, and for others who may also wish to know more about the overall scene of agriculture in China. As this preface is being written, visits to China by the above groups of people have been curtailed and further changes are taking place in the Chinese countryside. Some later book will have to record these, but the present signs are that some of the agricultural aid programmes will gradually be restarted, and some tourist programmes are recommencing. We hope that this book will be of value to all who wish to understand better the Chinese countryside and the many problems faced by the Chinese peasants.

In 1985 Professor Wang Maohua, Vice-President of the Beijing Agricultural Engineering University, visited the United Kingdom, and met, among others, Professor C. R. W. Spedding, Director of the Centre for Agricultural Strategy at the University of Reading. Professor Spedding told Professor Wang of a proposed book on the agriculture of China, to be prepared under the auspices of the Centre for Agricultural Strategy and to be published by Oxford University Press, and invited him to discuss the proposed book with Dr Peel. Professor Wang returned to China towards the end of the year and, after consultation with his colleagues, invited Professor Xu Guohua to be the Chinese editor of the book. Thus began a long and fruitful collaboration.

Many people have been involved in the preparation of this book and to all we offer our grateful thanks for their assistance. In particular we would like to thank Chen Xingchao, Song Lian, Wang Peizhong, Pang Zengshu, Zhao Dong, Liu Fengqin, Ke Baokang, and Wang Yilian who very kindly agreed to assist Xu Guohua with the translation of the Chinese text into English; Song Shenge, Sun Xuequan, Liu Qingshui, Wang Zenqi, Zhang Wei, and Chao Naiwen who provided information; Chen Hongchou who prepared some of the figures; and C. R. W. Spedding and Ruth Weiss who read and commented on the English text.

The British Council generously made it possible for Professor Xu to spend a month in England working with Dr Peel on the English text of the book, and as a result we are confident that the picture of Chinese agriculture which is now presented in English is clearer than might otherwise have been the case.

A number of changes which took place in China while the book was being prepared are not reflected in the text. These include the establishment of a new province of Hainan from a part of the old province of Guangdong, and the formation of the Ministry of Water Resources from part of the old Ministry of Water Resources and Electric Power. The name of the Ministry of Agriculture, Animal Husbandry and Fishery has been shortened to Ministry of Agriculture and although this change is mentioned in the text, the longer name has been retained for historical accuracy.

The tables in the book do not include information from Taiwan unless specifically stated.

Beijing and Reading　　　　　　　　　　　　　　　　　　　　Xu G. H.
March 1990　　　　　　　　　　　　　　　　　　　　　　　　L. J. P.

Contents

Contributors

Chen Dao, Professor, Department of Agricultural Economics, Beijing Agricultural University.

Chen Ren, Adviser, Institute of Modern Agriculture, Chinese Association of Agricultural Societies, Beijing.

Dong Kaichen, Professor, Library Division, Beijing Agricultural University.

Ke Baokang, Associate Professor, Department of Agricultural Machinery Design and Manufacture, Beijing Agricultural Engineering University.

Li Xiaochun, Adviser, Bureau of Education, Ministry of Agriculture, Beijing.

Mei Dunli, Senior Engineer, Chemical Engineering Planning and Design Institute, Ministry of Chemical Industries, Beijing.

Peel, L. J., Agricultural Consultant and Author, Reading, United Kingdom; Honorary Research Fellow, University of Reading.

Wang Weixin, Senior Engineer, Water Resources and Hydroelectric Power Planning and Design Institute, Ministry of Water Resources, Beijing.

Xie Lifeng, Assistant Statistician, Division of Forestry and Agriculture, State Statistical Bureau, Beijing.

Xu Guohua, Professor, Department of Irrigation and Drainage, Beijing Agricultural Engineering University.

Yang Shenghua, Professor, Department of Irrigation and Drainage, Beijing Agricultural Engineering University.

Yao Chaohui, Lecturer, Department of Agricultural Economics, Beijing Agricultural University.

Yu Yifan, Senior Engineer, Bureau of Animal Husbandry, Ministry of Agriculture, Beijing.

Zhang Lin, Engineer, Administration Division, Chinese Academy of Fishery Sciences, Beijing.

Zhang Qizong, Adviser, Division of Highway Construction, Ministry of Transportation and Communication, Beijing.

Zhu Zhaoling, Senior Engineer, Energy for Agricultural Uses Division, Institute of Agricultural Machinery, Nanjing.

List of Figures

List of Tables

1

Natural environment

XU GUOHUA

Geographic location

China is the third largest country in the world, after the USSR and Canada. It has a land area of 9.6 million km². This is 6.5 per cent of the total land area of the world, 40 times the size of Great Britain, or 91 per cent of the area of Europe as a whole. Located in the east of the continent of Eurasia, bordering the Pacific Ocean, China extends over 49° 16′ of latitude, from 4° 15′ N to 53° 31′ N. It thus lies mainly in the subtropical and temperate zones. The western border begins at 73° 43′ E, and the furthest eastern point is at 135° 5′ E, a distance of 65° 25′ of longitude. The northern part of the country is adjacent to the USSR and the People's Republic of Mongolia, the western part to Afghanistan and Pakistan, and the southern part to India, Nepal, Bhutan, Burma, Laos, and Vietnam. Japan is to the east of China across the Sea of Japan, and Korea lies to the north-east across the Yalu Jiang River (see Fig. 1.1).

Major natural geographic regions

China can be divided into three natural regions which differ significantly from each other in physical features. They are the Eastern Monsoon Region, the Xinjiang–Inner Mongolia Arid Region, and the Qinghai–Tibetan Plateau (or alpine) Region (Fig. 1.2).

Eastern Monsoon Region

The Eastern Monsoon Region is a part of the much larger southern and eastern Asia monsoon area where more than 50 per cent of the world's population lives. Forty-five per cent of China's land area is included within the region, and some 95 per cent of the total population of China lives there.

The wet monsoon in summer affects the climate markedly, producing a damp climate throughout the region. The natural vegetation is chiefly forest and partly forest–grassland. The western part of the region is a plateau at an altitude of about 1000 m, while most of the land in the eastern part lies below

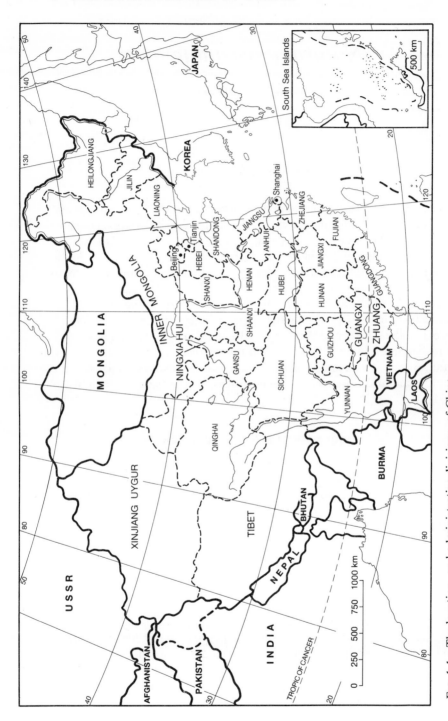

FIG. 1.1   The location, and administrative divisions, of China.
Source: Cartographic Publishing House (1984). *Map of the People's Republic of China*. Cartographic Publishing House, Beijing.

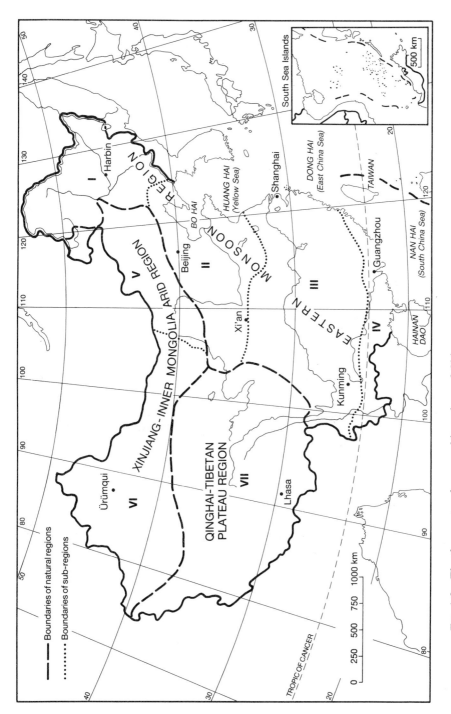

FIG. 1.2   The three natural geographic regions of China.
Source: Editorial Board (1984). Introduction, *The Natural Geography of China*. Science Press, Beijing.

500 m in a vast expanse of plains. Surface water replenishment is mainly from rainfall, with a certain amount of groundwater. In the Quaternary Glacial Period, glaciation in this region was not as extensive as in Europe and North America, therefore, the biota in this area suffered less and a broad species range developed. Irrespective of time, this region was, is, and always will be the major agricultural land of the country.

The Eastern Monsoon Region can be subdivided into four sub-regions on the basis of temperature differences due to latitude (see Fig. 1.2):

I.   Temperate humid/semi-humid North-East China;
II.   Warm temperate humid/semi-humid North China;
III. Subtropical humid central and South China;
IV.   Tropical humid South China.

The first two sub-regions tend to decrease in humidity with their distance from the sea in the east.

Xinjiang–Inner Mongolia Arid Region

The Xinjiang–Inner Mongolia Arid Region is part of the vast expanse of the central Eurasian desert. It covers 30 per cent of China's total area while containing only 4 per cent of the population. As a result of uplift in later geographic times, most of this region now consists of the Xinjiang–Inner Mongolian Plateau which lies at an altitude of about 1000 m, but within the plateau there are smaller areas where the land rises abruptly to form mountain ranges.

The climate throughout the region is characterized by its dryness, which increases steadily from east to west, although in the extreme west there are a few local exceptions to this. The dryness is due to the great distance from the coast, and to the surrounding mountains, which mean that the wet monsoon only has poor accessibility to the area. The available water on the plains comes from surface flows in the mountains. These flows are replenished by rainfall and snow thaw — the annual rainfall in the mountains is much greater than on the plains. The region has been desiccated gradually ever since the Mesozoic era and both the fauna and flora have many fewer species than in the Eastern Monsoon Region of China.

The region can be subdivided into two sub-regions on the basis of increasing aridity from east to west as shown in Fig. 1.2:

V.   Temperate steppe of Inner Mongolia;
VI.   Xinjiang temperate and warm temperate desert.

Qinghai–Tibetan Plateau Region

The Qinghai–Tibetan Plateau is the highest and largest plateau in the world and is known as the 'roof of the world'. This region constitutes 25 per cent of the country's territory, though less than one per cent of the country's people live there. Most of the region rises 4000–5000 m above sea-level and a number of peaks stand as high as 7000–8000 m. There are, however, a few lower lying areas in valleys in the south-east of the region and there is the Qaidam Pendi Basin in the north. The elevation leads to a very thin atmosphere, with widely distributed glaciers, intense exposure to solar radiation, and strong winds. Vegetation is sparse and, in general, stunted. There has been hardly any impact of human activity on the landscape, except in strips of the small valleys in the south-east and in the Qaidam Pendi Basin in the north.

The formation of the Qinghai–Tibetan Plateau

An important factor in determining the present existence of the three distinct geographic regions in China was the formation of the Qinghai–Tibetan Plateau. Uplifting of this plateau began in the Mesozoic era, in the Tertiary period. Before this uplifting, China was basically a low flat peneplain and part of the present plateau was even submerged in the sea. By Neogene, in the Miocene or Pliocene, the eventual plateau area was elevated to 1000 m or so above sea level. By the end of the Pliocene, a strong elevation made it rise to 3000 m. Then a violent uplift occurred early in the Holocene which raised the plateau further to its present height of more than 4000 m.

The rise of the Qinghai–Tibetan Plateau is best interpreted by the tectonic plate theory. The Indian plate which used to exist in the south maritime space gradually moved north and finally collided with the Eurasian plate. This collision and subsequent subduction of the Indian plate underneath the Eurasian plate, and its continued northward thrust, made the region of contact, the Qinghai–Tibetan Plateau, keep rising to the tremendous height of today. It is possible that the uplifting process is still in progress.

The rise of the Qinghai–Tibetan Plateau not only produced a unique climate and landscape on the plateau itself but also imposed important effects on the climate and landscape of the surrounding areas. Firstly, the plateau serves to bar the north-bound air current coming from the Indian ocean. Today the south-west monsoon has to travel through gaps and canyons to move north, and is unable to go very far. Even if a fraction of the south-west monsoon happens to get through the gaps and canyons, the moisture carried by the wind will be trapped on the south slope and precipitated in the form of snow and rain. This makes it almost impossible for the moist air current to reach the Xinjiang–Inner Mongolia Arid Region, hence the unusually low atmospheric moisture and rainfall there.

Secondly, unlike other regions of high altitude, the immensity of the area and its height tend to make its atmosphere a source of cold in winter and of heat in summer. During winter the regional atmosphere becomes a cold high-pressure atmosphere, and the surrounding free atmosphere at the same altitude becomes a depression. This gives rise to a north-east wind over the plain to the east of the plateau which reinforces the north-east monsoon, due to the distribution pattern of the land and sea. This not only leads to a low temperature in the east of China but also aggravates the aridity in north-west China. In summer, a warm low pressure appears over the plateau and reduces the subtropical high pressure at this latitude. The rainfall of a large area in the east of China is greatly increased by this warm low pressure, which makes the south-east and south-west monsoons there even more powerful.

Thirdly, the height of the Qinghai–Tibetan Plateau also causes the prevailing northern air current in the Northern Hemisphere in winter to diverge into two branches around the plateau and then to converge to its east and continue to move eastwards. This partition process moves the westerly wind belt several degrees of latitude to the south. When the partitioned air flow subsequently converges a trough is formed; the cold air behind the trough keeps moving southward and results in frequent southbound cold waves. In summer, because the prevailing westerly wind belt moves north, the southern branch disappears. However, the south-west monsoon then makes a northbound detour and often forms low-pressure eddy currents which move eastward, resulting in heavy rains and showers. This is the main source of rainfall in the Yangtze (or Chang Jiang) Valley and its vicinity in East China.

Topography

A distinct general feature of China's topography is its declination from west to east. The general picture is one of three terraces, or three major areas, remarkably different in altitude (Fig. 1.3). The Qinghai–Tibetan Plateau in the south-west of China is the highest terrace. To its east a 'Y' shaped area is the second highest terrace. The eastern boundary of this terrace is different from the western boundary of the central Eastern Monsoon Region of the three major natural geographic regions, in that it goes further east. In other words, approximately one-third of the western part of the central Eastern Monsoon Region is included in this terrace. More specifically, this terrace consists of the vast expanse of the Xinjiang–Inner Mongolian Plateau in the west, the Loess Plateau in the centre, and the Yunnan–Guizhou Plateau in the south. In general, the terrace lies 1000–2000 m above sea level, which is considerably lower than the Qinghai–Tibetan Plateau and moderately higher than the lowest terrace to its east. It should be noted that there are a few basins in this terrace, including the Sichuan Basin, standing only 250–700 m above sea level, and the even lower Turfan Basin which is 155 m below sea level. This is the lowest place in the country.

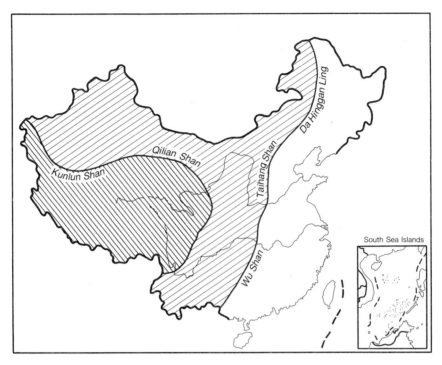

FIG. 1.3 The three topographical terraces of China.
Source: Xi Chengfan *et al.* (1984). *Fundamentals of Physical Regionalization of China*. Science Press, Beijing.

The lowest terrace lies in the eastern part of the country, where the plain of North-East China and the plain of the middle and lower reaches of the Yangtze River stretch successively from north to south. The elevation of this lowest terrace is below 200 m, and in many places it is below 50 m. To the south of the Yangtze River lies a large array of hills intermixed with numerous small plains. The hills are mostly less than 500 m in height, though some peaks rise to more than 1000 m. The exception to this are the 62 peaks in Taiwan which stand higher than 3000 m.

Generally speaking, China is a country of many mountains and many plateaux; this is especially so in western China. Topographically, the mountainous land comprises 33 per cent of the total area of China, the plateau land 26 per cent, the hilly land 10 per cent, and the basins 19 per cent — leaving a mere 12 per cent of the total area as plains. If the land of China is classified on the basis of elevation, 19 per cent is found to be above 5000 m; 18 per cent between 2000 and 5000 m, 28 per cent between 1000 and 2000 m and 16 per cent is below 500 m.

Climate

As discussed above, a monsoon climate prevails in the east of the country, which is part of the monsoon region of southern and eastern Asia under the powerful influence of the wet monsoon. The Qinghai–Tibetan Plateau, however, is too high, and the vast territory in North-West China too far away from the sea, to come into contact with the sea gales, and hence characteristically cold upland or arid climates are found. It is necessary to bear this picture in mind when looking at the climate of China.

Temperature

The temperature in the Qinghai–Tibetan Plateau is comparatively low because of its altitude. In other parts of China, the temperature decreases with the increase in latitude, which is a tendency easily understood. However, this tendency seems to vary in different seasons; in particular it is very noticeable in the cold season but less so in the warm season.

Figure 1.4 shows the mean temperature in the coldest month in China. In this figure, the isothermal lines in East China run almost parallel with the lines of latitude. The tendency for higher latitudes to be associated with lower temperatures is very pronounced.

Figure 1.5 shows the mean temperature in the warmest month. In this figure the tendency mentioned above can still be seen, but it is less marked. In fact, in the east, temperatures of 25°C can be experienced at 40° latitude in the warmest month yet the temperature is only 2.25 °C higher in the warmest month at 20° latitude. In South-East China practically no difference in temperature is observed over a large area.

From Figs. 1.4 and 1.5, it can be concluded that in winter there is a large difference in temperature between the north and the south, while in summer it is equally warm almost everywhere apart from on the Qinghai–Tibetan Plateau. For agriculture, it is ideal for the climate to be warm in summer but not so warm in winter. This will be discussed later.

The temperature in various places in China, in contrast with that in various other places in the world, is worth noting. For example, Huma county, in the province of Heilongjiang, has a latitude nearly the same as that of London. The average temperature of −27.8°C in January compares with an average temperature of 3.7°C in London, a difference of about 30°C. Again, Tianjin is at the same latitude as Lisbon, yet the average temperature in January in Tianjin is −4.1°C and that in Lisbon 9.2°C, a difference of 13°C. This is because the powerful cold winter winds in China move a long way south. In summer, however, the reverse is true, as the temperature in many places in China is much higher than in other places at the same latitude in the world. This is because China, backing onto the world's largest

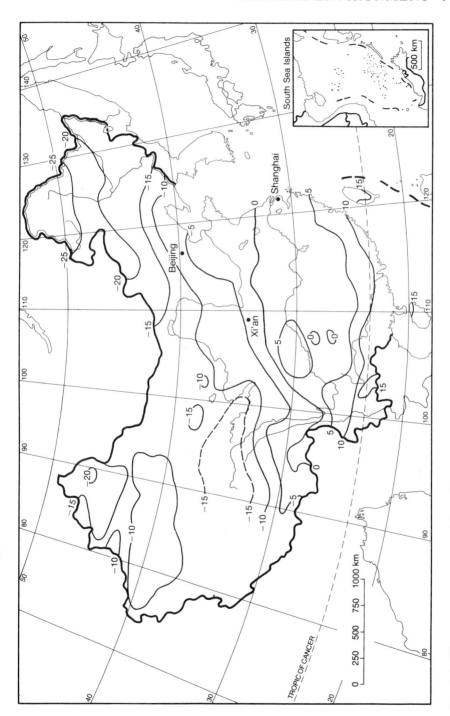

FIG. 1.4 The mean temperature of the coldest month (°C).
Source: Xi Chengfan, Qiu Baojian, Zhang Junmin, and Liu Donglai (1984). *Fundamentals of Physical Regionalization of China*. Science Press, Beijing.

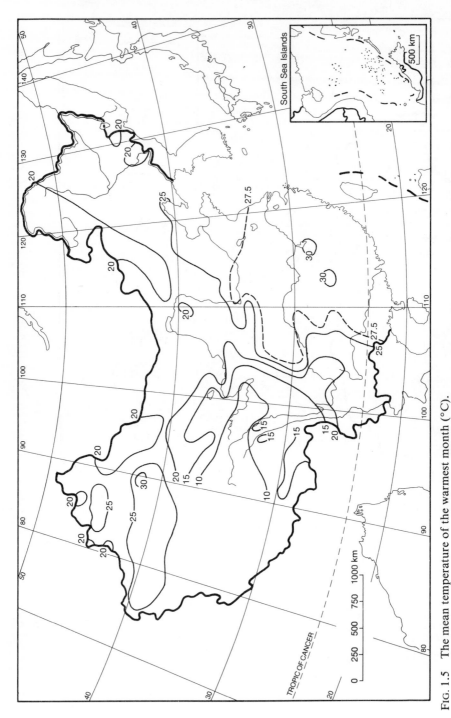

FIG. 1.5 The mean temperature of the warmest month (°C).
Source: Xi Chengfan, Qiu Baojian, Zhang Junmin, and Liu Donglai (1984). *Fundamentals of Physical Regionalization of China*. Science Press, Beijing.

continent and facing the world's largest ocean, has a continental climate. The great difference between cold season and warm season temperatures in China compared with the smaller differences in other parts of the world are a feature of this continental climate.

As well as the seasonal difference in temperature, there is a day and night difference. The day and night difference is not significant in the south where it hardly exceeds a few degrees Celsius. In North China, the day and night difference can be 10°C or more, and in North-West China it can be as great as 20–30°C — as the old saying goes: 'Wrapped in fur in the morning and draped in gauze at noon; enjoying a water melon while nursing a brazier'.

This is the general view of the climate of China. A few words will now be added about the influence of the climate on the agriculture of China.

The mean temperature in the coldest month, shown in Fig. 1.4, determines whether or not winter crops are able to grow, while the mean minimum temperature (Fig. 1.6) indicates whether or not the crops will be able to survive. The influence of cold weather is a dominating factor in Chinese agriculture and the distribution pattern of the crops is determined by it. The 0°C isotherm in Fig. 1.6 runs, in general, along the Nan Ling Mountain Range; the – 10°C isotherm roughly overlaps the Qin Ling Ridge–Huai He River line, and the – 25°C isotherm runs approximately along the line of the Great Wall. The boundaries of a number of China's cropping systems are based on these isotherms (see Fig. 1.7).

Frost-free periods

An adequate temperature is needed for the growth of crops, and for most crops 0°C is a critical temperature at which growth ceases. The frost-free period is the period during which the ground temperature is above 0°C, and the length of this period is a good indicator of whether it will be possible to cultivate a particular crop. The frost-free period in northern North-East China is only 100 days, and it is quite difficult to draw up an extensive farming plan for this area. To the south lies a large part of North-East China and Inner Mongolia, the Gansu Corridor and north Xinjiang, where the frost-free period lasts 100–150 days. Annual crops may be grown here, but in order to warrant harvesting, a few early-maturing crops such as broom corn millet and early maturing maize are preferred. Further south lies the North China Plain and a large part of Shaanxi and Shanxi provinces, where the frost-free period lasts 180–220 days. This allows for three crops to be grown in two years, or two crops in one year in a few locations. Still further south are the Yangtze Valley, where the frost-free period lasts for 220–280 days a year, and the Sichuan Basin, where the frost-free period lasts for more than 300 days. In these areas three crops can be grown in a year. In the southernmost part of China, including the areas south of the Nan Ling Mountain Range and the Yunnan–Guizhou Plateau, the frost-free period is well in excess of 300 days and crops can be grown throughout the year.

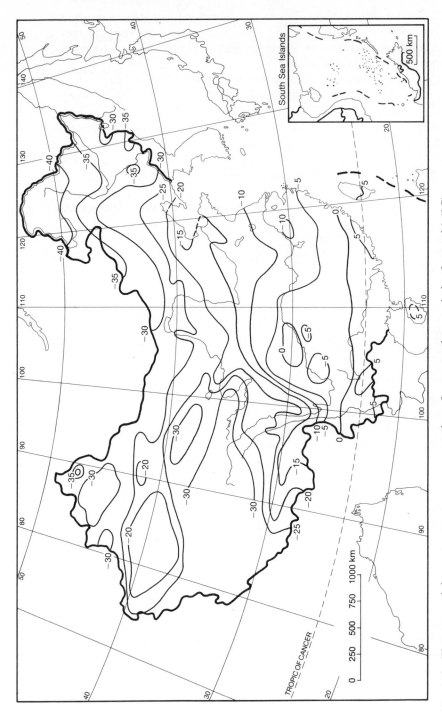

FIG. 1.6   The mean minimum temperature over a number of years (various periods of records) (°C).
Source: Xi Chengfan, Qiu Baojian, Zhang Junmin, and Liu Donglai (1984). *Fundamentals of Physical Regionalization of China.* Science Press, Beijing.

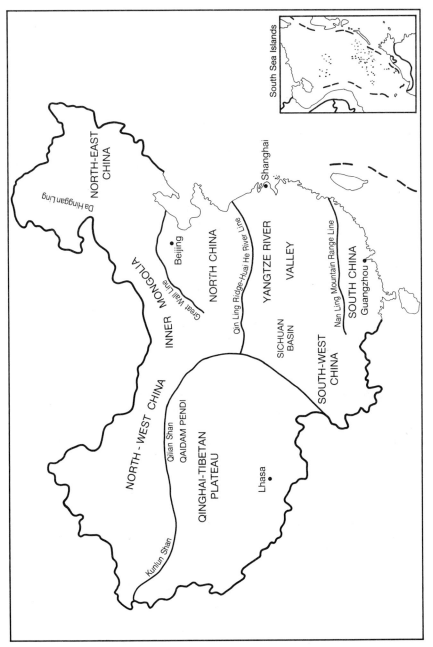

FIG. 1.7 Isotherm lines and geographical areas of agricultural significance.

Cumulative temperature and consecutive days above 10°C
A mean daily temperature of 10°C is another important criterion in determining the growth of crops. At mean temperatures above 10°C, thermophilic crops may grow, and ordinary crops can grow actively. The number of consecutive days with mean daily temperatures of at least 10°C may therefore be regarded as the length of the active growth period, and the cumulative mean temperature for days with such temperatures may be used as a measure of the solar energy available during the active growth period.

The first month during which the mean daily temperature passes through 10°C with some regularity is:

- January in places to the south of the Tropic of Cancer;
- January or February in places lying between 25° N and the Tropic of Cancer;
- February or March around the Qin Ling Ridge–Huai He River line;
- April in North China, Shaanxi, Gansu, and south Xinjiang;
- May, or in a few localities June, in North-East China, the greater part of Inner Mongolia, and north Xinjiang.

The last month during which the mean daily temperature is above 10°C is:

- September, or in some localities August, in North-East China, the greater part of Inner Mongolia, and north Xinjiang;
- October in North China, Shaanxi, Gansu, and south Xinjiang;
- December in places lying between 25° N and the Tropic of Cancer;
- January in places south of the Tropic of Cancer.

The number of consecutive days when the mean temperature is equal to or greater than 10°C is shown in Fig. 1.8 for various parts of China.

Figure 1.9 shows the annual cumulative mean temperature for the days with mean temperatures of at least 10°C. This is the sum of the mean daily temperatures in degrees Celsius for the consecutive days during which the mean temperature is 10°C or more. It can be seen that the cumulative temperature in North-East China, Inner Mongolia, and a large part of Xinjiang is below 3000°C; in North China it is 3000–4500°C; in the middle and lower Yangtze Valley it is 4500–6000°C; and further in the south it is even greater. In China, only wheat, millet, and sorghum can be grown in an area with a cumulative temperature of 2000–3000°C; winter wheat and other ordinary crops can be grown in places where the cumulative temperature is over 3000°C; thermophilic crops such as cotton require a cumulative temperature of over 4000°C; while in areas where the cumulative temperature is between 4500 and 6000°C two crops of rice can be grown in a year, and tropical crops of economic importance can also be grown. In areas with cumulative temperatures above 7000°C, various tropical crops are cultivated widely.

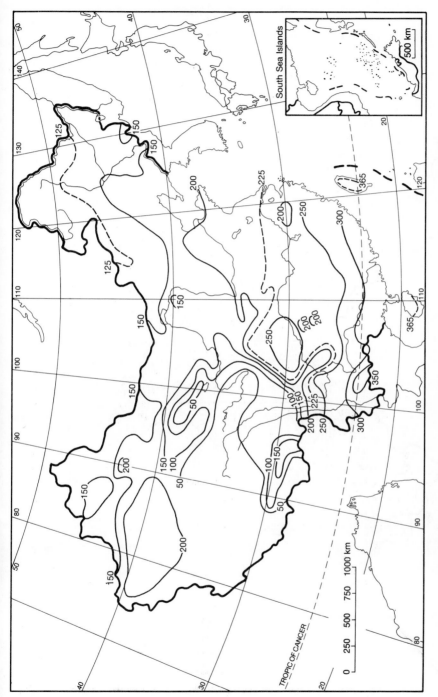

FIG. 1.8 The number of consecutive days per year with a mean daily temperature equal to or greater than 10°C.
Source: Xi Chengfan, Qiu Baojian, Zhang Junmin, and Liu Donglai (1984). *Fundamentals of Physical Regionalization of China*. Science Press, Beijing.

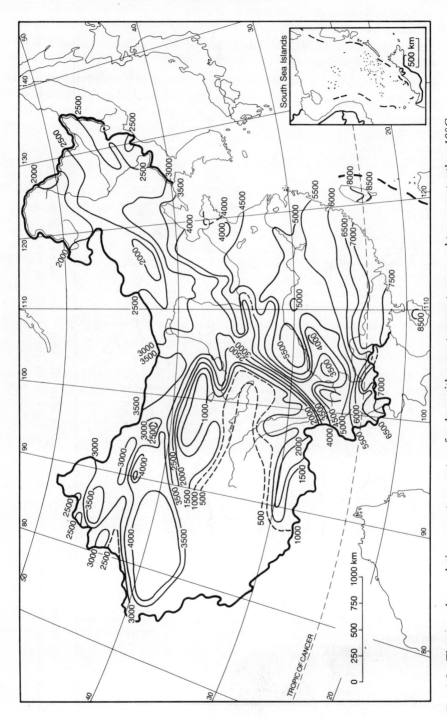

FIG. 1.9   The annual cumulative mean temperature for days with mean temperatures equal to or greater than 10°C.
Source: Xi Chengfan, Qiu Baojian, Zhang Junmin, and Liu Donglai (1984). *Fundamentals of Physical Regionalization of China*. Science Press, Beijing.

Rainfall

Annual rainfall

The mean annual rainfall (Fig. 1.10) is highest in the south-east coastal region, and from there it decreases across the country to the north-west, although in the extreme north-west it increases again over the mountains around the Ili He Valley where the air currents from the Arctic and Atlantic Oceans are able to penetrate. The annual rainfall in most of the south-east is 1500–2000 mm; in the middle and lower Yangtze Valley 1200–1400 mm; in places along the Qin Ling Ridge–Huai He River line 750–1000 mm; in North China 500–750 mm; in the southern part of North-East China more than 500 mm; in the western part of North-East China, most of Inner Mongolia, and the north-western part of the Loess Plateau, less than 500 mm and in parts less than 200 mm; in north-west Gansu and a large part of Xinjiang, less than 100 mm and in parts less than 50 mm; but in places along the Ili He Valley, 250–500 mm.

As shown in Fig. 1.10, the 400 mm isohyet begins in the north at the western foot of the Da Hinggan Ling Ridge, then runs south and turns west through the north-western part of the Loess Plateau, ending in the Qinghai–Tibetan Plateau. This isohyet runs in much the same direction as the previously mentioned boundary separating the Eastern Monsoon Region from the Xinjiang–Inner Mongolia Arid Region. Since the wet monsoon in summer rarely goes much beyond this boundary it has little or no effect there.

Seasonal distribution of rainfall

The seasonal distribution of the rainfall in most parts of China shows a concentration of the rains when the wet monsoon prevails, except in a few places where the distribution is comparatively uniform throughout the seasons. The eastern half of China to the east of the Yunnan–Guizhou Plateau is under the influence of the south-east Pacific monsoon, which usually arrives in South China in March and dominates the weather there in April and May. By the middle of June it reaches the middle and lower Yangtze Valley and, toward mid-July, turns north to the North China Plain and North-East China. It becomes dominant there in late July and early August, and its influence reaches southern Inner Mongolia and most of North-East China. In late August or early September it retreats rapidly to the south and about one month later the south-east monsoon leaves China. Correspondingly, South China becomes rainy in April, the Yangtze Valley early in June, North China early in July and the rainy season ends in August.

As a result of this pattern the rainy season lasts longer in the southern part of China, usually from spring until August, and there is a comparatively uniform distribution of the rainfall. In general, spring and summer each

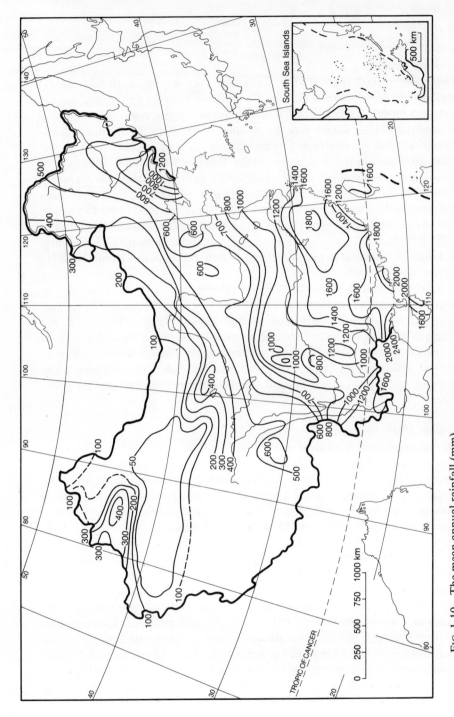

FIG. 1.10   The mean annual rainfall (mm).
Source: Xi Chengfan *et al.* (1984). *Fundamentals of Physical Regionalization of China*. Science Press, Beijing.

receive one-third of the annual rainfall, and autumn and winter share the remaining third. By contrast, the northern part of China has a short rainy season and well defined wet and dry seasons. Often one-half of the year's rainfall, or even two-thirds in some places, falls in the summer. The rainfall in the winter accounts for the smallest proportion, at less than 5 per cent of the total, while the rainfall in the spring is 10–15 per cent, and in autumn 15–20 per cent. Serious spring droughts are therefore common in northern China.

In the south-west Yunnan–Guizhou Plateau, the Sichuan Basin, and the south-east Qinghai–Tibetan Plateau, the rainy season starts when the south-west monsoon from the Indian ocean arrives, late in May, and lasts until October. The rainfall during these five months may contribute 80–90 per cent of the annual total. In the north-west arid region, where the annual rainfall is slight and a few showers may provide the total rainfall, the distribution of the rainfall is even less uniform. In China, the most uniformly distributed rainfall occurs in the Ili He Valley and in the northern part of the Altai Shan Mountains, both in the extreme north-west. Here the rainfall is evenly distributed throughout the four seasons.

Rainfall variability
Most of the rainfall in China is brought by the summer monsoon, but the dates at which the monsoon reaches and then retreats from a given place are not fixed. This means that the duration of the monsoon at any one place varies from year to year, and hence the variations in annual and seasonal rainfall are usually quite large.

The greatest variation in annual rainfall occurs in the north-west arid region, where it is always greater than 30 per cent, and in many places is up to 50 per cent. The variations in Inner Mongolia and most of North China are smaller than in the north-west, being in the range of 20–40 per cent. The variation in southern North-East China is below 15 per cent, but in the south-east coastal regions the variation increases to 15–20 per cent because of the influence of typhoons. The smallest variation in annual rainfall is in South-West China where it is less than 10 per cent as a result of the influence of the comparatively steady south-west monsoon.

A distinct feature of the climate of China is that the rainy season and the warm season coincide, apart from a few minor exceptions. This is quite different from the climate in many parts of Europe, especially where the Mediterranean climate dominates. It is worth noting, from an agricultural point of view, that the coincidence of the rainy and warm seasons is favourable: crops grow luxuriantly in the warm season with plenty of water available, while less rain in the cold season avoids the problems caused by thaw that occur in certain parts of Europe. However, over-concentrated rainfall may not only cause insufficient utilization of rainwater, but also result in detrimental rainstorms which give rise to soil erosion, floods, and waterlogging. Furthermore, scanty rainfall in winter is often the cause of drought. In China's history, many disastrous droughts and floods may be related to the non-uniform seasonal distribution of the rainfall.

The seasonal variability in rainfall may be as high as 100 per cent in spring in the north-west arid region, 50–80 per cent in North China and the South China coastal regions, and less than 30 per cent in the mid and lower Yangtze Valley. In summer, variability is still greatest in the north-west arid region, but in the mid and lower Yangtze Valley the variability increases to 60 per cent, while in North and North-East China it decreases to 30–50 per cent. South-West China has the lowest variability in summer rainfall at less than 30 per cent. In autumn, both East and South China have high rates of variability in rainfall of about 70 per cent, but in the Sichuan Basin the variability is only about 30 per cent. In winter the variability in rainfall is usually high. In North China it may go as high as 80–100 per cent, and in the mid and lower Yangtze Valley it is 50–60 per cent.

Rainfall intensity

Finally, a few comments on the intensity of rainfall may be useful. Throughout China, rainstorms are frequent in summer. In the south-east humid region, recorded daily rainfall of up to 100 mm or even 200 mm is not unusual. Rainstorms of such intensities do not necessarily make up a large proportion of the total annual rainfall in regions south of the Yangtze, where the rainfall is both large in quantity and frequent in occurrence. However, the torrential rain may be of an intensity which produces floods and soil erosion. In North China, where there is a limited annual rainfall, these torrential rains make up a considerable fraction of the annual rainfall, and in some cases a few such storms in summer may amount to 80 per cent of the total. Therefore, in North China, though the total rainfall is lower, floods and waterlogging are frequent, and serious soil erosion occurs. In the north-west arid region rain is scarce, and a few minor concentrated falls still account for a large proportion of the annual rainfall. The water resulting from rainstorms, however, cannot be efficiently utilized in agriculture unless it falls in places well equipped with water conservation facilities. A considerable proportion of the annual rainfall is, therefore, not necessarily beneficial to agriculture in China's semi-arid and semi-humid regions. This is worth noting when rainfall is considered in relation to China's agriculture.

In the Yangtze Valley and to the south of it, the rainfall is generally adequate in quantity, and its seasonal distribution is moderately uniform. Occasionally, concentrated falls or long periods between rains in summer may cause floods or droughts. North China does not suffer from a lack of rainfall but this tends to be concentrated into the summer period, resulting in drought in spring and waterlogging in autumn. Regions to the west and north of North China are obviously confronted with a shortage of water.

Distribution pattern of aridity

Figure 1.11 shows the distribution of aridity in China, where the aridity K is expressed as

$$K = 0.16 \, \Sigma t/r$$

where t is the sum of the daily mean temperatures (the cumulative temperature) for the days when the mean temperature is at least 10°C, r is the sum of the rainfall in such days, and 0.16 is a coefficient from the assumption that along the line joining the Qin Ling Ridge and the Huai He River the annual evaporation is approximately in equilibrium with the annual precipitation.

It can be concluded from the figure that the value of K increases in a direction from the south-east to the north-west. Based on the magnitude of K, China can be divided into four regions, namely:

(1) the humid region: $K < 1.0$ covering 32.0 per cent of the total area;
(2) the semi-humid region: $K = 1.0–1.5$ covering 14.5 per cent of the total area;
(3) the semi-arid region: $K = 1.5–2.0$ covering 21.7 per cent of the total area;
(4) the arid region: $K > 2.0$ covering 31.8 per cent of the total area.

The northern boundary of the humid region is the line joining the Qin Ling Ridge and Huai He River, the other three regions are located in the northern part of China and are situated successively in the direction from south-east to north-west.

Typhoons[1]

Another factor affecting China's weather which is worth mentioning is the typhoon. Typhoons invariably come into being at low latitudes over the Pacific Ocean and then move west or north-west. Some arrive in China on the south-east coast; in the 30 years from 1948 to 1978 276 typhoons reached China, an average of 9.2 typhoons per year. Of these, 155 were ordinary typhoons (average maximum speed 17.2–32.6 m/s), and 121 were strong typhoons (average maximum speed greater than 32.6 m/s).

The typhoons reach the Chinese coast most frequently in Guangdong (48.2 per cent), and the frequency decreases from there northwards. Only 6.2 per cent of the 276 typhoons recorded arrived on the Chinese mainland to the north of Shanghai. The typhoons usually arrive during July–September (79.7 per cent), though a few arrive in Guangdong later than October. Typhoons are usually harmful because of their very high speed and their

1. Since November 1988 the international system for classifying typhoons has been used in China, but the description of typhoons in this book is based on the classification of typhoons under the old system.

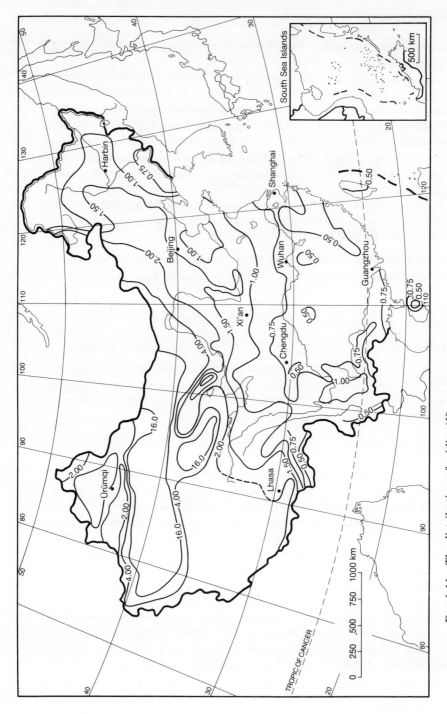

Fig. 1.11   The distribution of aridity (K).
Source: Editorial Board (1984). Introduction, *The Natural Geography of China*. Science Press, Beijing.

super-saturated water content. However, they do bring rain, and sometimes ease droughts, to the benefit of agriculture.

Rivers

There are countless rivers in China. More than 50 000 rivers have a drainage area greater than 100 km², and some 1500 rivers have a drainage area in excess of 1000 km². The rivers fall into two categories: the externally draining rivers that drain into the sea, and the internally draining rivers that flow into inland lakes or disappear into a desert or salt marsh. The external drainage systems occur mainly in the eastern regions where the monsoon climate is found, together with a few rivers in the eastern Qinghai–Tibetan Plateau. The internal drainage systems occur in the other parts of the country, although there are a few inland rivers in the plain of north-eastern China, and a river of external drainage in farthest northern Xinjiang. The basins of the externally draining rivers account for 64 per cent of the country's territory and 96 per cent of the water supply, while those of the internally draining rivers account for 36 per cent of the territory and a mere 4 per cent of the water supply. This shows the importance of the former. Eighty eight per cent of the externally draining rivers flow into the Pacific Ocean, and 10 per cent into the Indian Ocean; there is one river, the Irtysh River, which flows into the Arctic Ocean.

Geographic distribution of river basins

The best known rivers in China are the Yellow River (Huang He) and the Yangtze River (Chang Jiang). The basin of the former is the womb of the Chinese nation, and the basin of the latter is where the most productive land lies. They are the two longest rivers in the country, though they differ greatly in the amount of water they carry. The discharge of the Yangtze River amounts to 38 per cent of the total discharge of all the rivers in the country, whereas that of the Yellow River is a mere 2 per cent (Fig. 1.12).

The first five lines in Table 1.1 include all the externally draining river basins in northern China, down to and including the Huai He River Basin. These river basins account for 43 per cent of the area of all the externally-draining river basins in China, but their combined runoff only amounts to 13 per cent of the runoff from all such river basins. If we add the basins of the four internally draining systems, which are all in northern China, to the five externally draining river basins mentioned above, then it can be seen that all the river basins in northern China have an area amounting to 63 per cent of the land area of China, but the runoff within this area is only 17 per cent of the country's total.

On the other hand, the externally draining river basins in southern China, including the Yangtze River Basin, comprise only 57 per cent of the area of all externally-draining river basins, and only 37 per cent of the total area

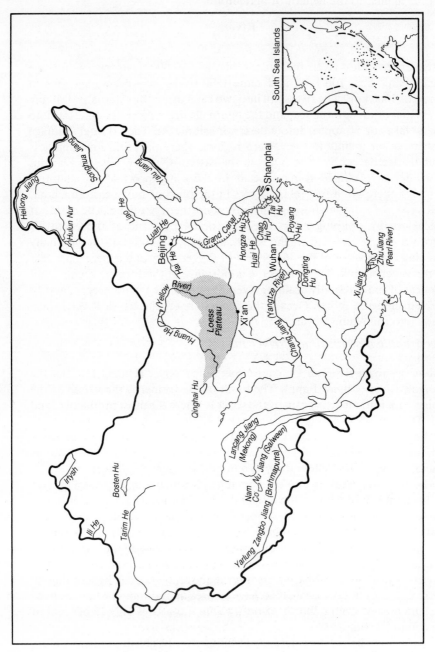

FIG. 1.12   The major rivers and lakes of China.

TABLE 1.1 The relative annual runoff, and mean depth of annual runoff, from the river basins of China

	Area (%)	Annual runoff (%)	Mean depth of annual runoff (mm)
Externally-draining river basins			
In North-East China	12.15	6.66	148
In North China	3.32	1.09	89
Huang He (Yellow) River Basin	7.84	2.21	76
Huai He, Mid-Grand Canal, Yi He, Shu He, and Shandong Peninsula River Basins	3.40	2.30	183
Irtysh River Basin	0.53	0.41	212
Chang Jiang (Yangtze) River Basin	18.83	37.66	542
In Zhejiang and Fujian Provinces	2.22	7.70	942
Of Zhu Jiang (Pearl) River and rivers in Guangdong and Guangxi Provinces	5.76	17.18	807
In South-West China	4.25	8.31	529
In Tibet	4.75	8.72	498
In Taiwan and Hainan Dao Islands	0.71	3.41	1302
All external drainage	63.76	95.65	406
Internally-draining river basins			
In Gansu and Xinjiang Provinces	21.77	2.73	34
In Qinghai Province and Tibet	10.55	1.47	38
In Inner Mongolia	3.42	0.10	8
In North-East China	0.50	0.05	25
All internal drainage	36.24	4.35	32
Total	100.00	100.00	

Source: Based on Xi Chengfan, Qiu Baojian, Zhang Junmin and Liu Donglai (1984). *Fundamentals of Physical Regionalization of China*. Science Press, Beijing. (In Chinese)

of the country; but the runoff from these basins amounts to 87 per cent of that from all externally draining basins, and to 83 per cent of the total for the country. In other words, rivers in southern China have a far greater discharge than those in northern China.

The mean depth of annual runoff over the surface area of the river basins is also given in Table 1.1. In the internally draining basins this does not exceed 38 mm, while in the Yangtze River Basin, and river basins further south, it is above 500 mm. Needless to say, the pattern of runoff is closely related to the rainfall distribution. Water resources south of the Yangtze River are plentiful, those in North-East and North China are insufficient, while in North-West China they are severely limited. Further details of the runoff from the major river basins in relation to population and cultivated area are given in Table 1.2. This table highlights the very much greater water resources available per hectare of cultivated land in the Yangtze and Pearl River Basins compared with the other major river basins, and the very high level of water resources that these two basins provide per head of the local population. The latter is undoubtedly a matter of importance to the future development of industry as well as agriculture.

Seasonal distribution of river flow

Since the major source of water flow in China is rainfall, the runoff reaches its maximum during the summer rainy season. The seasonal distribution of flow can be summarized as follows.

In winter (from December to February), there is little rainfall throughout the country, and the frost that prevails in the north drastically reduces river flow. For example, surface flow in the farthest northern part of North-East China diminishes to 1 per cent of the annual flow; that in other places in the north to about 2 per cent; south of the Yangtze River it decreases to 10 per cent; but in north-east Taiwan to only 25 per cent.

In spring (from March to May), a greater difference in surface flow from river to river is evident. Hilly regions south of the Yangtze River are watered by more than 30 per cent of the annual runoff. Ice-packed regions in the north of North-East China, in the east of Inner Mongolia, and in the north of Xinjiang receive 20–25 per cent of the annual runoff because of snow thaw. Regions in North China which are not ice-packed and which have little spring rain, receive only 6–8 per cent, and the rivers may even dry up, while regions in central and southern Yunnan receive less than 10 per cent of the annual runoff because of the delayed arrival of the south-west monsoon.

In summer (from June to August), the heaviest rainfall, as well as the year's largest surface flow, occurs everywhere in China. This is the time when almost all rivers are in their peak flow. For instance, rivers in the northern part of China run with more than 50 per cent of their annual flow, and in North China, Inner Mongolia, and southern North-West China the flow can amount to two-thirds or more of the year's total. In South-West China, with the arrival of the south-west monsoon, river flow can be up to 50–60 per cent of the year's total. However, in regions south of the Yangtze River, a comparatively smaller proportion of 35–40 per cent is observed, as a result of the northerly movement of the rainy zone.

In autumn (from September to November), there is also a considerable difference in surface flow from river to river. In South-West China, owing to the delayed evacuation of the south-west monsoon, a plentiful autumn rainfall leads to river flows of 30–45 per cent of the annual total. In the junction area of Shaanxi, Sichuan, Gansu, and Hubei provinces there is also

TABLE 1.2 .The annual runoff from the major river basins of China

River basin	Area (10⁴ km²)	Annual runoff (10⁸ m³)	(%)	Area of cultivated land (10⁶ ha)	Mean runoff per ha of cultivated land (m³)	Population (10⁸)	Mean runoff per person (m³)
Country total	960.0	26 000	100	100.0	26 000	10.00	2600
Songhua Jiang	52.8	759	2.9	11.7	6487	0.46	1650
Liao He	23.2	151	0.6	4.5	3356	0.29	521
Hai He	31.9	283	1.1	11.3	2504	0.92	308
Huang He (Yellow)	75.2	560	2.2	13.1	4275	0.82	683
Huai He	26.2	530	2.0	12.5	4240	1.25	424
Chang Jiang (Yangtze)	180.7	10 000	38.5	24.0	41 667	3.46	2890
Zhu Jiang (Pearl)	41.5	3070	11.8	4.4	69 773	0.74	4149

Source: Based on Xi Chengfan, Qiu Baojian, Zhang Junmin and Liu Donglai (1984). *Fundamentals of Physical Regionalization of China.* Science Press, Beijing. (In Chinese)

copious rainfall in autumn, contributing 35–40 per cent of the annual runoff, while in the North and North-East China Plains the rivers run with 25–30 per cent of the annual flow. Hainan Dao Island is exposed to typhoons, and the river flow in the autumn can be as high as 50 per cent of the annual flow.

It can be concluded from the above that, in general, there is only a slight seasonal change in flow for the Yangtze River and for other rivers south of the Yangtze. In South-West China river flow is less in winter and spring than in summer and autumn. For rivers in North and North-East China, the flow in winter and spring is usually very small, but the flow in rivers in the extreme north is comparatively larger in spring because of the snow thaw; in summer, the flow is much more concentrated. A plentiful water supply in summer is certainly a benefit to agriculture, yet a water flow that is excessively concentrated into one season causes difficulties in its utilization at the time of maximum flow, and a shortage of water in other seasons. The decline in river flow is not a major problem in China in winter, but in spring it becomes harmful when the crops begin to grow. By the end of spring (around May), when water is badly needed and surface flow in North-East and North China is slight, it can become the limiting factor for the growth of crops.

Coefficient of variation and yearly variation of runoff from river basins

The coefficient of variation C_v is a measure of the difference in a river's flow from year to year over a period of a number of years. The expression for C_v is given as follows:

$$C_v = \{\Sigma(K - 1)^2/(n - 1)\}^{0.5}$$

where n = the number of consecutive years in the time period, and K = the ratio of one year's river flow to the mean value for river flow in all years in the time period.

The greater the value of C_v, the greater the variation in river flow from year to year.

The values for C_v for China's major rivers are: Yangtze River, 0.12–0.15; Yellow River, 0.45; Huai He River, 0.55–0.65; Hai He River, 0.60–0.75; Songhua Jiang River, 0.41; Xi Jiang River (the main tributary of the Zhu Jiang or Pearl River), 0.23; Lancang Jiang River, 0.17; and Yarlung Zangbo Jiang River, 0.20. Of the aforementioned rivers, the Yellow River, Huai He River, and Hai He River are in North China, the Songhua Jiang River is in North-East China, the Yangtze and Xi Jiang Rivers are in southern China, and the Yarlung Zangbo Jiang River and Lancang Jiang River are in South-West China. A comparison of the C_v values shows that rivers in South and South-West China have rather low values and the Yangtze the lowest;

whereas rivers in North China have the highest C_v values. This is a typical distribution pattern of the C_v values of rivers in China.

The ratio of maximum to minimum annual flow in rivers varies geographically in the same way as the variation in C_v, except that the differences are more marked. In general, the ratio of maximum to minimum flow for rivers south of the Yangtze River is less than 2.0, whereas the ratio for rivers in North China is usually 5.0–10.0. The highest ratio may be as high as 50 — for example, for the Tuhai He and Majia He Rivers in Shandong, and for a few rivers in Inner Mongolia.

If the wet years and dry years occur alternately in a regular manner there are few problems; but if either wet or dry years occur continuously for a number of years then the situation becomes serious, especially in North China. Table 1.3 shows that rivers in northern China have the highest number of consecutive years either of inundation or of low water, and also the largest variation from the average for the level of river flow during both the inundation and the low water periods. The differences are most conspicuous of all in the Songhua Jiang River and the Yellow River.

Water resources other than rivers

Lakes

There are 2800 lakes in China with a surface area greater than 1 km², with a total surface area of 80 000 km². Details of some of the more important lakes are given in Table 1.4. Lakes Poyang Hu, Dongting Hu, Tai Hu, Hongze Hu, and Chao Hu are freshwater lakes in the area of external drainage; the remainder of the lakes listed in Table 1.4 are inland lakes and most of them are salt.

Lakes in the area of external drainage are replenished by rivers, and they invariably play a role in adjusting river flow. For example, there are five tributaries of the Yangtze River which pour water into Lake Poyang Hu. The lake water is then discharged through Hukou, the outlet from the lake, into the Yangtze. If flood peaks arise in the tributaries, the lake can be used to reduce peak flow by 15–30 per cent. This is particularly important when catastrophic floods occur. A tremendous flood in the Yangtze River in 1954 poured a flow of 45 800 m³/s into the lake (17 June 1954), but the lake's maximum flood discharge into the Yangtze (20 June 1954) was only 22 400 m³/s. Another example is Lake Dongting Hu, into which four tributaries of the Yangtze pour their water. The water then flows out of the lake through Tiaoxuankou, into the Yangtze. Again, a catastrophic flood in 1954 poured a huge flow of water at 69 000 m³/s into Lake Dongting Hu on 30 July; but the lake's maximum flood discharge into the Yangtze (2 August 1954) was 41 600 m³/s.

TABLE 1.3 The number of years in succession during which the runoff from various river basins was greater than average, and less than average

Location	River basin	Hydrologic station	Years observed	Longest wet period observed (years)	Runoff in the longest wet period as a percentage of the average	Longest dry period observed (years)	Runoff in the longest dry period as a percentage of the average
North-East China	Songhua Jiang	Harbin	78	8	132	13	60
North China	Luan He	Luan Xian	37	4	126	4	69
	Huang He (Yellow)	Shaanxian	56	9	118	11	75
	Huai He	Bongbu	51	3	199	6	72
South China	Chang Jiang (Yangtze)	Yichang	90	4	113	6	91
	Min Jiang	Zhuqi	38	3	130	7	80
	Xi Jiang	Wuzhou	32	3	118	6	82
	Lancang Jiang	Yunjinghong	17	2	125	4	91
	Nu Jiang (Salween)	Dabajie	16	6	109	3	87
	Yarlung Zangbo Jiang	Nuxia	17	3	120	5	87

Source: Xi Chengfan, Qiu Baojian, Zhang Junmin and Liu Donglai (1984). *Fundamentals of Physical Regionalization of China*. Science Press, Beijing. (In Chinese)

TABLE 1.4   Important lakes in China

Lake	Location	Area (km²)	Altitude (m)	Maximum depth (m)	Capacity (10⁸ m³)
Qinghai Hu	Qinghai	4583	3195.0	32.8	1050.0
Poyang Hu	Jiangxi	3583	21.0	16.0	248.9
Dongting Hu	Hunan	2820	34.5	30.8	188.0
Tai Hu	Jiangsu	2420	3.0	4.8	48.7
Hulun Nur	Inner Mongolia	2315	545.5	8.0	131.3
Hongze Hu	Jiangsu	2069	12.5	5.5	31.3
Nam Co	Tibet	1940	4718.0	–	–
Bosten Hu	Xinjiang	1019	1048.0	15.7	99.0
Chao Hu	Anhui	820	10.0	5.0	36.0

Source: Xi Chengfan, Qiu Baojian, Zhang Junmin and Liu Donglai (1984).
Fundamentals of Physical Regionalization of China. Science Press, Beijing. (In Chinese)

Glaciers

A number of glaciers are found in the high mountainous regions of the Qinghai–Tibetan Plateau, with an estimated area of 57 000 km², and a total aqueous reserve of 2940×10^9 m³. These glaciers are, as a rule, the fountainheads of rivers. The annual melt water supply from the glaciers has been estimated at 50.46×10^9 m³, and this replenishes the rivers with fresh water. It should be noted that in some of the extremely arid regions in northern Xinjiang and on the brink of Qaidam Pendi Basin in the north of the Qinghai–Tibetan Plateau, melt water from the glaciers becomes the basic condition for the formation of various oases. It is necessary to study the availability of local water supplies of this kind when planning the development of agriculture in this area.

Groundwater

There are large reserves of groundwater in the plains of eastern China. The plains in North-East China are usually rich in groundwater resources of high quality except in the coastal regions. The North China Plain is also rich in groundwater, but its quality is variable. Parts of the plain adjacent to the mountains, and near the south of the plain, have water of high quality, but the water in the north of the plain is saline (1–2 g salt/l), and that in the coastal regions of the plain may contain 3–5 g of salt per litre. The plains in southern China are also rich in groundwater of high quality. In North-West China there are only a few places where groundwater can be found in quantity. Unfortunately, not much of it is suitable for irrigation because of its salinity.

The total groundwater reserve is estimated at 800×10^9 m³/a, or 31 per cent of the country's surface water supply, but the distribution is uneven. Three-eighths of China's groundwater spreads over a vast area in North China, while five-eighths is in the south of the country. Only a small proportion of the water in the north is suitable for use in agriculture and husbandry. Already, in North China, the groundwater in certain regions has been over-extracted and the shortage of groundwater has become a problem.

Soils

The main soils and their distribution

In a large country like China the great variation in physical features across the regions produces a wide variety of soils. According to the classification of the China Soil Science Society, the soils in China fall into 46 great soil groups, but within the scope of this book, a simplified description only can be given, as follows.

With the exception of the Qinghai–Tibetan Plateau, there are two major aspects of the climate of China that should be noted. Firstly, the aridity in northern China tends to increase from east to west. Secondly, both the temperature and the humidity in eastern China tend to increase from north to south. Accordingly, there are two series of zonal soils in China that reflect these climatic features.

The north series is found in the northern part of China, extending from North-East China via Inner Mongolia, Ningxia, and Gansu to Xinjiang. Thus, this series extends over the whole of North-East China and the arid region of Inner Mongolia and Xinjiang. The main great soil groups in this series are, from east to west in succession: black soil — chernozem — chestnut soil — brown soil — sierozem — desert soils.

The east series is found in the eastern part of China, extending from North-East China via North China, the Huai He and Yangtze River Basins, and the region to the south of the Yangtze River, to South China. That is, it extends over the whole of the Eastern Monsoon Region. The major great soil groups in the east series are, from north to south in succession: chernozem and black soil — dark loessial soil — drab soil and brown earth — yellow-brown earth — yellow soil and red soil — laterite.

There is also a third series of great soil group soils, in the Qinghai–Tibetan Plateau, called the Qinghai–Tibetan series. In this region there is an increase in aridity from south-east to north-west, and the major great soil groups from the south-east to the north-west successively are: alpine meadow soils — alpine steppe soil — alpine desert soils. All these are special great soil groups which exist in extremely high plateaux or mountains.

The three series of zonal soils and their distribution can be shown

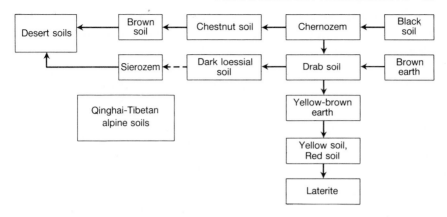

FIG. 1.13 A schematic diagram of the distribution of the main zonal soils of China.

schematically as in Fig. 1.13. The black soil and chernozem are distributed over North-East China; the chestnut soil over eastern Inner Mongolia; the brown soil and sierozem over western Inner Mongolia, Ningxia, and Gansu; and the desert soils mainly over Xinjiang. In the east series the dark loessial soil, drab soil, and brown earth are distributed over North China; the yellow-brown earth over the region lying between the Yangtze and Huai He Rivers; the yellow soil and red soil over the region south of the Yangtze River; and the laterite over South China.

The distribution pattern can be simplified further as shown in Fig. 1.14. In this simplified diagram the North-West desert soils include the brown soil, sierozem, and desert soils; the North-East dark soils include the black soil, chernozem, and chestnut soil whose colour is distinctly darker than that of the other soils; the North China yellow soils include the brown earth, drab soil, and dark loessial soil which are basically formed from loess parent material and have a yellow colour; the southern China red soils include the

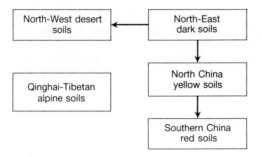

FIG. 1.14 A simplified schematic diagram of the distribution of the main zonal soils of China.

yellow soil, red soil, and laterite which have a distinct reddish colour. The rough diagram shown in Fig. 1.14 cannot be considered a soil map, but it does provide a straightforward and symbolic outline for the reader.

Important soil forming processes relating to soil properties

It is useful to look briefly at the characteristics and properties of these soils in terms of some of the natural processes that occur during soil formation.

The north series

In the north series of soils there are two natural processes which are worth noticing.

The first is the accumulation and decomposition of organic matter. If accumulation prevails over decomposition, the organic matter content in the soil will be high. This is the case with the black soil and chernozem regions in North-East China, where in summer it is warm and wet (400–500 mm rainfall in the chernozem region, and 500–700 mm in the black soil region) and the luxuriant growth of plants provides a plentiful supply of organic matter. In addition, the long, freezing winters limit bacterial activity so that the decomposition of organic matter is inhibited, and the organic matter content of the soil is thereby kept high. For example, the organic matter content in the organic horizons of both the chernozem and black soil is 3–7 per cent, but the thickness of the organic horizon of the black soil is as much as 70 cm, whereas in the chernozem it is only 30–40 cm. Toward the west the climatic conditions in the chestnut soil region are similar to those in the former two regions, except that the rainfall is less — only 300–400 mm. The region is, therefore, less densely covered with vegetation, and the organic matter content of the chestnut soil, at 1–4 per cent, is lower than that of the chernozem or black soil. The organic horizon is also thinner and measures only about 20 cm. The further west the soil extends, the less rainfall it receives, the less the vegetative cover, and the lower the organic matter content. The organic matter content of both the brown soil and the sierozem is less than 1 per cent, while that of the desert soils is still less, almost zero. Correspondingly the organic horizon grows thinner and becomes unnoticeable.

The second kind of natural process in the north series of soils that is worth mentioning is that of eluviation and illuviation (eluviation is the movement of material in solution or suspension through the soil, and illuviation the deposition of such material in a soil horizon). During eluviation, salt, clay, or other materials move downward, and may be leached out of the soil body if there is a strong eluviation; hence the lack of certain components in the surface soil or through the soil profile. If, however, the eluviation is weak, the matter will be trapped in certain sections of the soil profile, producing stratified deposits. Eluviation in the black soil in the eastern part of the north series is strong, resulting in the absence of CaCO$_3$ in the soil profile, and producing a slightly acidic soil (pH 5.5–6.5). By comparison, eluviation in the chernozem is somewhat weaker and CaCO$_3$ generally can be found. Sometimes, a calcic horizon can be seen in the lower or middle part of the profile, although monovalent cations will have been leached out, and the soil reaction is neutral or slightly alkaline (pH 6.5–8.0). Further west is the chestnut soil region where eluviation is weaker still, and illuviation becomes the dominating factor: the soil is rich in lime and the calcic horizon is plainly visible. Still further west is the region where brown soil, sierozem, and desert soils are distributed; here eluviation is even more negligible while illuviation is much more important.

The east series

The processes that the east soil series undergo and which are worth noticing are: eluviation and illuviation, the accumulation and decomposition of organic matter, and the allitic process.

Eluviation is strong in the south because of the heavy rainfall which penetrates through the soil so frequently that not only the monovalent but also the divalent metallic cations are leached out of the soil profile completely. Thus soils to the south of the Qin Ling Ridge–Huai He River line are acidic, or strongly acidic, with unsaturated bases, and the exchangeable cations are dominated by hydrogen and aluminium ions. In contrast, soils to the north of this line are fundamentally neutral or alkaline, base saturated, and with calcium and magnesium ions dominating the exchangeable cations. Among these latter soils there is a difference between those distributed in the east and those in the west. Those in the east are exposed to more rainfall and stronger eluviation, so there is hardly any lime left in the profile of the black soil and brown earth. There is also little lime left in the surface of the drab soil, although some is found in the middle and lower horizons. The dark loessial soil is rich in lime throughout the profile. A certain amount of downward movement of clay particles also occurs in these soils.

The accumulation and decomposition of organic matter is different in the east series of soils compared to that in the north series. Soils in the north of eastern China are subjected to rains and warmth in summer, but the period suitable for the growth of vegetation is obviously shorter than that in the south. This leads to a smaller resource of plant material in the north than in the south. On the other hand, the long, cold winter in the north keeps the decomposition of organic matter at a lower level than that in the south. The combined effect of these two factors leads to uncertainty when a comparison is attempted between the organic matter contents of the northern and southern soils. What is certain, though, is that a high organic matter content is always associated with the preservation of vegetation and, conversely, the destruction of vegetation and frequent tillage of the soil invariably decrease the organic matter content. As these farmlands of China have been in service for many years, and have been ploughed innumerable times, the organic matter is at a very low level, probably generally only 1.0–2.0 per cent, and with some of it at only 0.7–1.2 per cent.

The allitic process occurs only to the south of the Yangtze River where, because of the warmth and high rainfall, silicate clay minerals in the soil tend to be strongly hydrolysed. The basic materials, together with the silicate ions released by hydrolysis, are leached out of the soil profile leaving a preponderance of aluminium and iron compounds in the clay fraction. The allitic process is usually expressed in terms of the silica–sesquioxide ratio: the ratio for the red soil in China is 2.0–2.2, and for the yellow soil, 2.5–2.7. These soils are red or reddish yellow in appearance. The yellow soil is distributed over the Yunnan–Guizhou Plateau, where the humid atmosphere causes the iron and aluminium oxides to be hydrated and acquire a yellowish colour. It should also be added that the intense weathering caused by the warm and humid atmosphere in regions south of the Yangtze River makes the texture of both the red soil and the yellow soil more clayey. Under these conditions kaolinite becomes the major component of the clay mineral fraction, resulting in a low base exchange capacity, generally lower than 10 mEq per 100 g of dry soil.

Loess as a soil-forming parent material

In addition to the processes mentioned above, the importance of the Loess Plateau and of the loess as a parent material is also worth noting. The Loess Plateau extends over about 300 000 km², and here dark loessial soils and similar soils have developed. Because of frequent torrential rains in North China in summer, along with the high susceptibility of the Loess Plateau to erosion, all rivers that flow through the plateau have a high silt content. The most famous example is, of course, the Yellow River which has a silt content as high as 35 kg/m³, and an annual silt discharge of 1.6×10^9 tonnes, 90 per cent of which is carried from the Loess Plateau. As a result, most of the soils of the North China Plain are developed from loessial alluvium and have an unmistakable loess character — that is, a high content of silt, a general texture of silt loam, and a yellowish coloration.

Azonal paddy and saline soils

In addition to the zonal soils, there are also some azonal soils in China. The two most important of these are the saline soil and the paddy soil.

China's saline soil is distributed over flat land north of the Qin Ling Ridge–Huai He River line, mostly in the north-west arid region — with the exception of those saline areas caused by tidal inundations, which can occur even in the coastal belt of Taiwan and Guangdong where the rainfall is plentiful. Saline soil is a problem soil in China, as it is elsewhere in the world, but the country's arable land is so over-populated that even saline soil, which may be left to lie waste in other countries, has to be cultivated. The Chinese people are, therefore, experienced and successful in the reclamation and cultivation of saline soil.

Paddy soil is formed during the course of a long period of rice growing, which changes the original properties of the soil. It is distributed chiefly over the area to the south of the Yangtze River, and because of the differences in the environment from place to place, the fertility of this soil is variable.

Other problems of soil use in China, apart from those of saline soil, include the reclamation of red soil waste land, soil and water conservation on the Loess Plateau, control of aeolian sandy soil, reclamation of lime-concreted black soil, and amelioration of cold-water paddy soil.

Natural vegetation

The natural vegetation of China, compared with that of other countries, is rich in variety but small in quantity. China owes its richness in variety of natural vegetation to the under-development of glaciers during the period when glaciers were much more extensive in Europe. Ancient botanical species are, therefore, preserved in great numbers. It has been estimated that there are 301 families, 2980 genera, and 24 490 species of Spermatophyte in China, among which there are 2800 tree species, and all the Gymnosperms in the world except for one family. For example, there are 1400 species of trees and bushes in Hainan Dao Island, and 900 in Xishuang Banna; yet in Europe there are only 30 genera of broad-leaved trees, and seven genera and 18 species of conifer. The lack of quantity of natural vegetation is obviously associated with China's huge population and long history of cultivation and human activity. According to a nation-wide survey in the years 1973–76, there were 122×10^6 ha of forest in China, including planted afforestation areas, representing 12.7 per cent of the country's territory. This was equivalent to a per caput forest area of 0.13 ha; a per caput annual timber reserve of 9.0 m³; and a per caput annual timber share of 0.22 m³. These figures correspond to 12.6 per cent, 13.8 per cent, and 32.4 per cent of the world's average per caput figures respectively, and they are indeed small figures.

Forest

As mentioned earlier, China can be divided into three natural geographic regions. The Eastern Monsoon Region is generally favourable for forest, and nearly all the land grew forest before it was cultivated. The tree species here vary as the temperature and humidity change from high to low latitude. There are seven types of forest:

(1) Cold Temperate Coniferous Forest — in northern Heilongjiang. Typical tree species: Dahurian larch.

(2) Moderate Temperate Coniferous and Broad-Leaved Mixed Forest — in south-eastern North-East China. Typical tree species: Korean pine, maple, lime, and birch.

(3) Warm Temperate Deciduous Broad-Leaved Forest — in North China. Typical tree species: Mongolian oak, East Liaoning oak, oriental white oak, sawtooth oak, oriental oak, Japanese red pine, and Chinese pine.

(4) North Subtropical Evergreen and Deciduous Broad-Leaved Mixed Forest — in central China. Typical tree species: sawtooth oak, oriental oak, *Castanopsis sclerophylla*, and blue Japanese oak.

(5) Central Subtropical Evergreen Broad-Leaved Forest — in the Yangtze Valley. Typical tree species: blue Japanese oak and evergreen chinquapin; masson pine, Chinese fir, and bamboo are also present.

(6) South Subtropical Monsoon Evergreen Broad-Leaved Forest — in South China and Taiwan. Typical tree species: camphor, *Castanopsis sclerophylla*, evergreen chinquapin, banyan, and lily magnolia.

(7) Marginal and Central Tropical Rain Forest and Equatorial Coral Reef Evergreen Forest — on Hainan Dao Island and South China Sea Islands. Typical tree species: mulberry, myrtle, custard apple, Chinese soapberry, euphorbia, and palm.

It should be added that because the winter monsoon can reach far south and the summer monsoon can reach far north in China, the forest pattern may differ from that in other parts of the world. In particular, a definite contrast can be observed on opposite sides of a mountain ridge. For example, broad-leaved forest grows on the east slope of Lupan Shan Mountain where the summer monsoon reaches and the rainfall is 500 mm, while forest steppe vegetation covers the west slope where the summer monsoon is obstructed and the rainfall is only 390 mm. Another example is Wuzhi Shan Mountain (Five Finger Mountain, on Hainan Dao Island, Guangdong province) where tropical monsoon rain forest grows on the east windward slope, where the annual rainfall exceeds 2000 mm, whilst savannah type vegetation grows on the west lee slope where the rainfall is below 1000 mm.

In the western half of China forests are sparse. In the Xinjiang–Inner Mongolia Arid Region, there is only a small area of forest in the Ili He Valley in the north-west, where air currents from the Arctic and Atlantic Oceans penetrate, and the land can be watered by snow thaw from the high mountains; while in the Qinghai-Tibetan Plateau Region, it is only in the river valleys in a small part of the south-east that a few forests are found. All the forests in these areas are subalpine coniferous, and they consist, typically, of fir and spruce.

Grassland and desert

The main vegetation over the Xinjiang–Inner Mongolia Arid Region and the Qinghai–Tibetan Plateau Region is grassland. It can be subdivided as follows, according to the natural conditions:

1. Temperate Meadow Grassland — in the eastern-most part of Inner Mongolia, where the annual rainfall is 330–500 mm. This occupies the transitional area from grassland to forest. It consists of graminaceous, leguminous, rosaceous, and ranunculaceous species. The herbage grows well and the fresh fodder yield reaches 3000–3750 kg/ha. This is the best pasture in China.

2. Temperate Steppe — in central Inner Mongolia, the south-west of North-East China, the northern part of the Loess Plateau, and some areas of Ningxia, Gansu, Shaanxi, and Xinjiang. The annual rainfall is 300–400 mm. The pasture consists of graminaceous, composite, leguminous, and rosaceous species. Ground cover is 30–50 per cent and the fresh fodder yield 1500–2200 kg/ha, occasionally 3000–4500 kg/ha.

3. Temperate Desert Steppe, or Semi-Desert — in the centre and west of Inner Mongolia, and in some parts of other provinces or autonomous regions in North-West China. The annual rainfall is 150–250 mm and sandstorms are common. The vegetation consists mainly of xerophilous herbs and some xerophilous shrubs. Ground cover is 15–30 per cent, and the fresh fodder yield 400–1200 kg/ha.

4. Alpine Grassland — in areas above 4000 m in the Qinghai–Tibetan Plateau, above 3000 m in north-western Sichuan, and above the forest line in the Altai Shan, Tian Shan, and Qilian Shan Mountains. The herbage consists of composite, rosaceous, and ranunculaceous species, most of which are cold tolerant and drought tolerant. The pasture grows only to a low level, the yield is not high, and the growth period is short. There are few toxic species present however, and the herbage is palatable and its nutritional value is high.

5. Desert — in Xinjiang, west Inner Mongolia, and the north-western part of the Qinghai–Tibetan Plateau, where the annual rainfall is less than 200 mm, and in some places less than 50 mm. The vegetation in the desert is very sparse, apart from a few xerophilous shrubs, and the ground coverage is only 5–10 per cent.

There are also some small areas of grassland distributed in the Eastern Monsoon Region.

According to the statistical record, there are about 286×10^6 ha of various kinds of grassland in China, distributed in 227 stock raising or semi-stock raising counties and 'banners' over ten provinces and autonomous regions, such as Xinjiang and Inner Mongolia. This area of grassland occupies 29.6

per cent of the total area of the country. Of the 286×10^6 ha of grassland, 63 $\times 10^6$ ha are subject to severe natural conditions and/or a lack of access and so are unavailable for use. The remaining grassland, which amounts to 223 $\times 10^6$ ha or 23.4 per cent of the total land area, can be used for raising animals. However, most of this usable grassland lies in the arid Inner Mongolia–Xinjiang Plateau or the extreme alpine Qinghai–Tibetan Plateau, where grass can grow luxuriantly in a few places only. The fodder yield is quite low over large areas because of the severe environmental conditions.

Wildlife

As China was not affected seriously by the Early Quaternary glaciation, the wildlife species, like the plant species, are numerous. There are more than 2000 terrestrial vertebrate species in China — 10 per cent of the world's total. These include a number of the world's most treasured wildlife, such as the giant panda, Mongol wild ass, Mongol mustang, hump-nosed antelope, takin (the ox-antelope), wild camel, white-lipped deer, Manchu tiger, South China tiger, gibbon, golden snub-nosed monkey, red-crowned crane, red ibis, and mandarin duck. But China, with its enormous population and long history, has been subjected to man's activities wherever the natural conditions are not too harsh. In addition, the harsh conditions in the arid and alpine zones limit occupation of these areas to those wildlife species specially adapted to survive in them. These factors contribute to a lack in the quantity of wildlife, and some species have been so reduced in numbers that they are on the verge of extinction.

Fish and other aquatic resources

China has a fish resource of about 2100 species. Of these, 1300 are marine fish and they include 300 commercial species of which 60 are common species regularly caught. In the past, big yellow croaker (*Pseudesciaena erocea*), small yellow croaker (*P. polyocilis*), hairtail (*Trichiurus haumela*), and cuttle fish (*Sepia* spp.), the so called 'four main marine fishes', made up most of the yield. However, the catches of big yellow croaker and small yellow croaker have decreased rapidly for various reasons, and these species are no longer among the main commercial species. At present hairtail and trigger fish (*Cantherines* spp.) are the main species caught, and other commercial fish are mackerel, matreel, silvery pomfret, porgy, Chinese herring, and grey mullet. (The trigger fish is called rubber fish in Chinese as it has a thick, rubbery skin — it is not used for human food).

There are more than 200 species of marine shellfish in China. The main commercial species are oyster, mussel, bloody clam, razor clam, scallop, and abalone. Nearly 2000 species of algae are found on the Chinese coast, including about 60 species used commercially, though kelp and laver make up nearly all the harvest. The most important crustacean on the Chinese coast is the prawn (*Penaeus orientalis*), and both its output and value exceed those for any other crustacean. In recent years there has been a rapid development of prawn cultivation, and the output of cultivated prawns now exceeds that of caught wild prawns. Sea cucumbers are the most important echinoderms along the Chinese coast, and development of their artificial culture is now proceeding rapidly.

There are over 800 species of freshwater fish in China, of which more than 220 are commercial species. These include black carp, grass carp, silver carp, big-head carp, and crucian carp; other fish, such as kilsa herring, whitebait, dog salmon, rainbow trout, and eel are also prized. In addition there is a rapid development of other freshwater species such as crab, shrimp, turtle, rice-field eel, and pearl mussel. China's freshwater aquatic production is comparatively advanced, and the output ranks first in the world. Of this production over 80 per cent comes from fish farming.

References

The following texts are in Chinese:

Cheng Lu and Lu Xinxian (1984). *The Agricultural Geography of China.* Agricultural Publishing House, Beijing.

Editorial Board (1984). Introduction, *The Natural Geography of China.* Science Press, Beijing.

Institute of Geography, Chinese Academy of Sciences (1983). *The Natural Conditions and Natural Resources of Agriculture in China.* Agricultural Publishing House, Beijing.

Xi Chengfan, Qiu Baojian, Zhang Junmin, and Liu Donglai (1984). *Fundamentals of Physical Regionalization of China.* Science Press, Beijing.

The following text is in English:

Zhao Songqiao (1986). *Physical Geography of China.* Science Press (China) and John Wiley Inc. (U.S.A.), published jointly.

2

The historical and social background

DONG KAICHEN

The Chinese dynasties

The culture of China, in which the Han nationality predominates, originated in the middle reaches of the Yellow River valley. The ruins of a primitive community of 4000 BC, discovered in Banpo village, near Xi'an in Shaanxi province, provide evidence of this. These ruins, located on the terrace of Weishui tributary, occupy an area of 5000 m². In the centre, there are numerous remains of house foundations and cellars for grain storage, all of which are round or square in shape. The earthenware pots from the site even contained some intact millet and vegetable seeds. Other similar cultural relics of the Neolithic Age have been found at hundreds of other sites in the area.

The location of the cradle of Chinese civilization along the Yellow River and its tributaries is assumed to be closely related to the nature of the loessial soils which cover these areas. In the early period of historical development the friability and fertility of these soils made them suitable not only for primitive methods of cultivation but also for the digging of holes for house building, and this helped the formation of settlements of tribal communities in ancient China.

Up to about the twenty first century BC, tribes living in the middle and lower reaches of the Yellow River were merging into a single nation. Later generations called these people the 'Hua-Xia', and it was they who established the earliest Chinese dynasty (c.2000–1520 BC). It is said that Yu was chosen as the chief of the tribal alliance for his success in leading and organizing a campaign to tame a flood which threatened a vast area. Yu divided the whole area under his rule into nine administrative divisions and cast nine tripods of bronze to symbolize his unified leadership. It was also accepted that Yu could hand over the crown to his son, and a hereditary system and a feudalistic royal court gradually evolved.

The Xia kingdom lasted over 400 years, and was replaced by the Shang kingdom (c.1520–1030 BC) (see Table 2.1). The Shang were organized tribally, and originally lived along the lower reaches of the Yellow River, under the rule of the Xia kingdom before its replacement. After they had seized power through a succession of conquests the Shang expanded their

TABLE 2.1 The Chinese dynasties

Xia kingdom	c.2000–c.1520 BC
Shang kingdom	c.1520–c.1030 BC
Zhou dynasty	c.1030–221 BC
Early or Western Zhou period	c.1030–722 BC
Eastern Zhou period	
Chun Qiu (Spring and Autumn) period	722–480 BC
Zhan Guo (Warring States) period	480–221 BC
First unification	
Qin dynasty	221–207 BC
Han dynasty	202 BC–AD 220
Western Han	202 BC–AD 9
Xin interregnum	AD 9–23
Eastern Han	AD 25–220
First partition	
San Guo (Three Kingdoms) period	AD 221–265
Wei	AD 220–265
Shu Han	AD 221–264
Wu	AD 222–280
Second unification	
Jin dynasty	AD 265–420
Western Jin	AD 265–317
Eastern Jin	AD 317–420
Second partition	
Nan Bei Chao (Northern and Southern dynasties)	AD 386–581
Southern dynasties	
(Liu) Song dynasty	AD 420–479
Qi dynasty	AD 479–502
Liang dynasty	AD 502–557
Chen dynasty	AD 557–589
Northern dynasties	
Northern (Tuoba) Wei dynasty	AD 386–535
Western (Tuoba) Wei dynasty	AD 535–556
Eastern (Tuoba) Wei dynasty	AD 534–550
Northern Qi dynasty	AD 550–577
Northern Zhou (Xianbi) dynasty	AD 557–581
Third unification	
Sui dynasty	AD 581–618
Tang dynasty	AD 618–906
Third partition	
Wu Dai Shi Guo (Five Dynasties and Ten Kingdoms) (Later Liang, Later Tang (Turkic), Later Jin (Turkic),	AD 907–960

TABLE 2.1   (*Cont.*)

Later Han (Turkic), and Later Zhou dynasties)	
Liao (Qidan Tartar) dynasty	AD 907–1124
West Liao (Kara Qidan) dynasty	AD 1124–1211
Xi Xia (Tangut Tibetan) state	AD 986–1227
Fourth unification	
Song dynasty	AD 960–1279
Northern Song dynasty	AD 960–1126
Southern Song dynasty	AD 1127–1279
Jin (Jurchen Tartar) dynasty	AD 1115–1234
Yuan (Mongol) dynasty	AD 1260–1368
Ming dynasty	AD 1368–1644
Qing (Manchu) dynasty	AD 1644–1911
Republic of China	AD 1912–1949
People's Republic of China	AD 1949–

Source: Based on Bray, Francesca (1984). *Volume 6: Biology and Biological Technology, Part II: Agriculture.* In Joseph Needham (ed) *Science and Civilisation in China.* Cambridge University Press, Cambridge.

territory, not only dominating North China and the area bordering Inner Mongolia in the north, but crossing the Yangtze River into the south. After the founding of the Shang kingdom, its capital was moved six times during the 500 years or more of the kingdom's existence, though all sites were within the present Shandong and Henan provinces. Some scholars suggest that the main reason for transferring the capital was the need for fresh grazing land, as nomadic grazing formed a major part of the economy. Other scholars have suggested that the capital was moved in a search for fertile land, since the level of agricultural technology was quite low and reasonably high crop yields could not be sustained for very long in one place.

As well as the development of agriculture and handicrafts, trade eventually came into being, and cattle were brought into use to transport goods over the longer distances. Trade was extended east to the seashore and west to the continental interior, as shown by the presence of sea shells and jades found in the relics of the Shang kingdom. The relics that attract most interest, however, are the bronze wares such as food utensils, drinking vessels, arms, and tools for production — they are not only exquisite in shape but were also intended for a diversity of uses. Pictographic characters carved on cattle shoulder blades or tortoise shells record the forecast of good or bad omens of divination, indicating that the Shang kingdom was the first dynasty in China's history to possess a recorded history in written characters.

The Shang kingdom lasted some 500 years and was then succeeded by the

Zhou dynasty (1030–221 BC); this was the longest dynasty in Chinese history, lasting for about 800 years. However, it was only in the early stage that the central government could keep effective control over its whole territory. This early stage (before 722 BC) is called by historians the Western Zhou; its capital was situated near the present city of Xi'an. Later, the Zhou dynasty moved its capital to Luoyang, and this second period has been called the Eastern Zhou. This was a time of great turbulence and the period can be divided into two stages: the Chun Qiu (Spring and Autumn) (722–480 BC), and the Zhan Guo or Warring States (480–221 BC). Since there were always contradictions, struggles, and wars among the nobility who received enfeoffment from the Zhou dynasty, it was quite natural that their numbers should decrease while the strength of the remaining members of the nobility increased. Some of the preferred nobility became kings or even emperors in the period of the Warring States, and there was more frequent and more fierce fighting among the nobility for hegemony. However, the tendency to bring all powers under one rule was also becoming more and more apparent. There were seven major powers during the time of the Warring States; the Qin kingdom, situated in the western part of North China, finally wiped out the other six powers, and unified the whole of China. The king who led this process of unification became the first emperor of the whole of China in 221 BC; he was Qin Shi Huangdi.

The Chun Qiu and the Warring States periods formed an epoch in China's history in which both production and intellectual theories flourished. Innovations in agricultural production during this time included the introduction of the iron plough drawn by oxen, and the gradual appearance of an individual farming economy. The land had previously been cultivated communally using a wooden (or much more rarely bronze) plough and the peasants had paid nearly all of their produce to the rulers, who in turn had provided the peasants with part of their requirements. The improvement in the construction of the plough, through the introduction of iron, made it possible for one family to operate on its own. To meet the needs of these new developments the rulers came to approve private ownership of land and the buying and selling of land, and in some small kingdoms a system of taxing peasants according to the land that they held was introduced. The repeated exchange and sale of land facilitated the shaping and development of a landlord class. Land could also be acquired through enfeoffment from the feudalistic state, and the feudal landlord system (which differed from the European feudal system) originated in China during this period.

After the unification of China, Qin Shi Huangdi established an autocratic centralized monarchic system which lasted for some 2000 years and had a great influence on Chinese history. The Qin dynasty, however, did not last long (221–207 BC). The governors of the Qin dynasty constantly ordered the people to labour in constructing the Great Wall as a defence against the Xiongnu (Hun) tribe on the northern border, and in connecting the highways to maintain internal military links, during the hard times after the protracted warfare. Eventually the demands became so remorseless that the peasants rebelled and the Qin dynasty was overthrown.

In the post-war destitution, the newly established Han dynasty (202 BC–AD 220) adopted a policy of rehabilitation by reducing the corvée and taxes. By the time of Emperor Han Wu Di (140–87 BC) the national strength had been restored and somewhat strengthened, the central government had been consolidated, and the territory expanded. Through military conquests, and marriage arrangements in later years between the two royal families (Han and Hun), the northern nomadic Hun tribe was made subject to Han authority, thus removing threats from the northern border. In addition, the Han dynasty brought into China's territory the areas west of North China where some minority nationalities lived, and laid the foundation for communications between the states in middle Asia and China. Gradually, through a long period of intermingling, the former 'Hua-Xia' nationality merged with neighbouring nationalities to form a new nationality, the Han, and despite the future continual readjustment of inter-nationality relations the territory of China as a multinational country began to have a definite shape.

In the earlier stage of the Han dynasty (202 BC–AD 9) both the population and agricultural production were still concentrated in the middle and lower reaches of the Yellow River. The south remained an economically backward area with a sparse population in a vast territory. It was not until the later Han dynasty (AD 25–220) and the succeeding San Guo (Three Kingdoms) period (221–265) that the situation of the south began to change. Climatic conditions were better in the areas south of the Yangtze River than in North China, and there was greater potential for agricultural development. To exploit this land full of bushes and swamps required considerable labour and suitable techniques, however. In the later period of the Han dynasty there was continual warfare among the feudalistic, separatist war-lords in North China which forced a part of the population to migrate to the south. Furthermore, two of the Three Kingdoms of the San Guo period were located in the middle and lower reaches of the Yangtze River. In order to take in the immigrants and sustain the economy, exploitation of the land had to be speeded up, and this gave impetus to the development of agriculture in the south.

The dynasty which ended the struggle among the Three Kingdoms and unified the whole of China was the Western Jin (265–317), which had Luoyang as its capital. Not very long after its establishment, it was overthrown by the Hun from the north. The Jin dynasty, however, retained sovereignty in the southern part of the country and there became the Eastern Jin dynasty. The south and the north were, therefore, partitioned and confronted each other for 270 years. Later historians called this period Nan Bei Chao (Northern and Southern dynasties). The North went through some ten dynasties ruled by five nomadic tribes successively, and the South underwent the rise and decline of four dynasties following the fall of the Eastern Jin. It was not until AD 581 that the whole of China was re-unified by the Sui dynasty (581–618). Some members of the nomadic nationalities living in the northern border area moved into North China during the Nan Bei Chao; as a rule, those who moved furthest from their place of origin into the compact communities of the Han nationality were the most ready to change their habits to become feudal farmers with small parcels of land. As the differences in economic activities faded, the merging of nationalities accelerated, and this again carried within itself the possibility of re-unification.

In order to promote the political unification of the whole of China, the Sui dynasty excavated the Grand Canal to link the economies and cultures of the south and north. The Canal went from the Taihu Plain in the south to the Hebei Plain in the north, and had a branch to Luoyang from the middle reach of the Canal. One of it most important functions was to link up most of China's big rivers which usually flowed from west to east — these included the Yangtze River, the Huai He, the Yellow River, and the Hai He — thus facilitating the flow of traffic and transportation.

The Sui dynasty was succeeded by the famous Tang dynasty (618–906) which had its capital at Chang'an (now Xi'an). This was the most flourishing and prosperous age in China's history. Chang'an became not only the political centre of China but also the focus of economic and cultural interchange between Asian nations. Merchants of the Tang dynasty haunted the markets of some of the Asian nations, and businessmen from western Asian countries came to China for trade along the Silk Road.

At the end of the Sui dynasty many of the landlords had been killed or had gone into exile, and their lands were taken over and controlled by the Tang dynasty. Part of this land was then given to the peasants, according to their family size, and this provided an impetus for the development of agriculture. In the later years of the Tang dynasty annexation of land and exorbitant taxes again became rampant. Prompted by a serious drought in North China, the peasants joined in armed rebellion, and the Tang dynasty eventually collapsed in 906 under the peasant insurrection.

In the succeeding half-century five successive short dynasties came to power in the northern part of China (907–960), and ten smaller separatist war-lords seized power in the southern part of China. Chinese historians refer to this period as Wu Dai Shi Guo, which means five dynasties and ten kingdoms.

The Song dynasty (960–1279) was then established, and unified first the central part and then the southern part of China. However, there were still some political powers of the minority nationalities in the northern border areas. The unification achieved by the Song dynasty depended on the combination of labour from the north and materials from the south. Early in the middle of the Tang dynasty, after generations of development, the economic strength of the south had become equal to that of the north, and after its fall the small powers in the south tried hard to develop agriculture by constructing irrigation schemes and cultivating more land. Thus, at the beginning of the Song dynasty, although the present Kaifeng, in Henan province, was made the capital, and North China was maintained as the political and cultural centre, the key economic areas were to be found in the south-east part of China.

During the Song dynasty the social status of the peasants was somewhat improved in that their census registration was changed; each peasant was registered directly with the state instead of being attached to a landlord. From this time the idea of peasants as part of the property of the various landlords gradually disappeared, and the peasants became the people of the state, at least partly independent of the landlords. The relationship of the landlords to the peasants became one in which the landlords leased land and collected rent in kind.

The development of economic production during the Song dynasty brought about an advance in science and technology — the invention of gunpowder for use in weapons, the application of the compass to navigation, and the development of the movable, individual-character printing press were all accomplished in this period.

Compared with its economic and cultural achievements, the Song dynasty was weak in military strength. The Jin kingdom, established by a nomadic tribe in the north-eastern part of China, gradually invaded the Song territory, and finally captured the Song dynasty's capital in AD 1126. The Song dynasty then moved its capital to Ling'an (now Hangzhou, in Zhejiang province) and became the Southern Song dynasty. Henceforth, China was again split into two opposing parts. The Southern Song dynasty was compelled to carry out a policy favourable to economic development so as to accumulate sufficient strength to prevent further intrusion into the south by the Jin. Meanwhile the agriculture in North China underwent new changes from the middle of the Jin dynasty (AD 1115–1234). At that time, people of Han nationality made up at least two-thirds of the total population of the whole of China, and the population of peasants would have comprised not less than two-thirds of the total population, so the productive activities in the Jin kingdom were mainly carried on by peasants of Han nationality. Because the nobility of the Jin dynasty frequently took land from the peasants of Han nationality and reallocated it to their own subordinates, there were many disputes among the people of the Jin dynasty, which eventually resulted in a serious decline in agricultural production.

Until the thirteenth century the nomadic Mongols, from the Mongolian grasslands, had successively conquered a number of nations in Asia and Europe, and during the thirteenth century they established the Yuan dynasty in China (1260–1368). They named their capital Dadu (the great capital), at the place where there has long been the capital of China — it is now called Beijing.

At the beginning of the Yuan dynasty the agriculture of North China lay ravaged from long years of warfare; in addition the ruling people of Mongolia were prejudiced against cropping. They considered that the Han people, who were engaged in cropping, would not improve matters and went so far as to transform cultivated land into grassland, and gave orders that no horses or vehicles should be used to draw farm implements for cropping. However, under the influence of the highly developed agricultural economy of the Han peasants to the south, the Mongols were at last obliged to give up their backward nomadic economy and turn to agricultural exploitation of the land. The members of the Mongolian ruling class gradually became feudalistic landlords, and measures were taken to restore the damaged cropping system.

North China received most damage during the Mongolian war of conquest; southern China remained in a relatively peaceful situation with a better agriculture than in North China. After the establishment of the Yuan dynasty, over three million hectolitres of rice were sent annually from the south to the capital, to meet the daily requirements of the loyal members of the ruling class and the bureaucrats. It was for this purpose that the Yuan dynasty restored the Grand Canal, but its depth and width were not very great, and boats with heavy loads could not use it. Only some hundreds of thousands of hectolitres of rice could be transported along the canal each year, the remainder had to go by sea, despite the risk of loss via that route. After the beginning of the Yuan dynasty frequent improvements were made to the Grand Canal, and eventually its transporting capacity was increased to some three to four million hectolitres of rice per year. The Grand Canal then became the main channel for transporting grain from south to north, and continued as such until 1855, when the Yellow River changed its course and burst through the dykes and banks of the Grand Canal. This forced the re-adoption of sea transport, but before long a railway line was built and the more modern form of transport replaced the old. During the Yuan dynasty there was much traffic between China and other Mongolian kingdoms in mid-Asia and Europe. In general, China became better known to the outside world during this period, for example, through the famous travel notes written by Marco Polo.

The next dynasty was the Ming dynasty (1368–1644), again established by the Han nationality. About 100 years after the founding of the Ming dynasty the European armed merchantmen reached China. Traditional China, though shocked by this intrusion, continued with its economic life largely unchanged and with its trade restricted, until the middle of the Ming dynasty. From then on commercial activity developed rapidly. In areas south of the Yangtze River the modern employer–employee relationship even arose in the handicraft industry.

In the later period of the Ming dynasty, homeless people in the north rebelled against the corruption of the ruling class. Some high ranking officers recognized the crisis and tried to introduce reforms to retrieve the situation but failed. The insurrection became increasingly serious, and finally the rebel army captured Beijing. The last emperor of the Ming dynasty committed suicide on Jinshan behind the Forbidden City.

The last feudalistic Chinese dynasty was the Qing or Manchu dynasty (1644–1911). This was a kingdom of the Man nationality in the North-East region before it succeeded the Ming dynasty. As the insurrectionary army was capturing Beijing, the Manchurian army seized the opportunity to come through the Shanhaiguan gate (at the east end of the Great Wall), and from there it gradually pushed southward and eventually unified the whole of China.

At the beginning of the Qing dynasty, measures were taken to improve the relationships between the different nationalities in order to win their support. Of these the Han were the greatest in number and the Qing rulers reached a compromise with landlords of the Han nationality, and abolished some of the corrupt practices of the later Ming dynasty. They also endeavoured to stabilize public order and restore and develop the economy by such measures as exempting a family's new-born children from poll tax or merging the poll tax into the land tax. As a result there was a period of nearly 100 years of prosperity during the early part of the Qing dynasty. However, during these years of prosperity a new crisis arose. Apart from those circumstances which had commonly featured in the fall of previous dynasties, such as high concentration of land ownership, the corruption of officials, and insufficient financial resources, the Qing dynasty also had to face difficulties arising from the internal pressure of population growth and the external challenge set by western capitalist countries.

With the long period of peace and improving productivity the population of China increased from about 100 million at the beginning of the Qing dynasty to 400 million in the mid-nineteenth century. In comparison with the growth of population, the area of cultivated land was only increased from about 5 million qing (1 qing equals 16.47 acres or 6.67 ha) in the Ming dynasty to 7 million qing in the Qing dynasty. In order to overcome this problem of the population increasing more rapidly than the land resources could be expanded, the people had to raise the yield from the cultivated land, and some also migrated from the densely populated areas to the remote border areas.

When people in North China migrated they usually went to North-East China where the Man nationality came from, but the government of the Qing dynasty at first forbade them to do this. It later gradually relented, and as a result of the ensuing migration most of the suitable land in North-East China was cultivated during the years from the mid-nineteenth to the mid-twentieth centuries. Meanwhile, people living in the coastal provinces of southern China migrated to other places in South-East Asia, and a considerable number of people of Han nationality migrated from the densely populated eastern part of China to the south-west and the north-west where people of the minority nationalities lived.

In the Opium Wars of 1839–42 and 1856–8 British gunboats eventually opened the forbidden gate of the Qing dynasty. This made China open to trade with the outside world from its secluded self-contained economy, but also reduced China to a semi-colony of the foreign nation. Feudalism in the rural areas was heavily shaken by these events but was not destroyed and it continued to hinder the introduction of new social changes. Hard times followed, with parts of China's territory being lost to belligerent neighbours, and many people had great difficulty in obtaining a living. The founding of the People's Republic of China in 1949 ended this stage.

This sketch of Chinese history is included to provide an historical background to facilitate the understanding of Chinese agriculture. From this sketch we may note that there were two important factors which have deeply influenced the development of Chinese history and agriculture.

The first is that though the rise and decline of every dynasty has had its specific causes, a comparison of the dynasties reveals a similar underlying theme: the struggle for land. In pre-capitalist societies land was the main means of production, and in these societies the person who seized the land gained the power to allocate the wealth. In China this struggle was particularly apparent and fierce because arable land has nearly always been in comparatively short supply. Hence the struggle for land involved not only the landlords and the peasants, but also the nobility, the officers and the ordinary landlords of the ruling group; everyone wanted to seize more land. This led to perpetual social turbulence. It was a general rule that each new dynasty would adopt a policy aimed at fostering a balance between population and land, for at that time the population would have been reduced by the war leading to the founding of the dynasty. The land rent which the landlords took from the peasants, and the land tax which the government took from the landlords and also possibly from the peasants, were set at levels which the peasants were considered to be able to pay. The government did not, therefore, impose excessive taxes and levies on the peasants at the beginning of a new dynasty. However, with time the landlords and officers would try to take over peasants' or even public land for themselves. In addition, some of the more powerful landlords also evaded paying land taxes to the government. This not only made life very hard for the peasants, but it also left the government short of financial resources and incapable of sustaining its rule. Once the situation reached this stage, the only way out was to wait for the peasants' insurrection to bring a temporary settlement: the root cause of these struggles remained and was by no means eradicated.

The second underlying factor in Chinese history is that most of the feudal dynasties were established by the Han nationality. The Han effectively extended their rule to the remote borders but their life-style depended on cropping and this could never be extended into the grasslands of the interior. The minority nationalities, who lived a nomadic life, entered North China and southern China through peaceful migration or conquering war, and even established central government in China, but could not sustain their nomadic life-style in these areas for long. Although the conflicts between nationalities led to the partition of China many times, the tendency towards unification grew stronger after each period of partition. Historically, the term 'China' denotes not only the dynasties in the central part of China, but also the regional authorities and tribal organizations of the minority nationalities in the border areas. The unification of the whole country achieved by the Qing dynasty promoted the final formation and consolidation of a multinational China. However, the different life-styles and modes of production of the different nationalities still remain, and are still in the process of mixing and assimilation. The contrast of crop land and grassland shows clearly in space the changes in the rise and decline of the Han nationality who live on intensive cropping in a definite settlement, and of the minority nationalities who live a nomadic life.

The historical development of the territory and population of China

China is a multinational country consisting of 56 nationalities with a total population of over 1000 million people.

'China' is now an abbreviation for the People's Republic of China, but the meaning and geographical range represented by this word has varied at different times. 'China' has been used to represent the central part and inland, where the 'Hua-Xia' tribes lived; from the Han dynasty on it was used to represent the country of the dynasty established in the central part (mainly North China). However, the dynasties established by the minority nationalities, such as the Mongol and Manchu, did not use the word China but the name of the dynasty to represent their country. For example, the Qing dynasty used Da Qing (the great Qing) to represent their country in concluding treaties with foreign countries. After the Republic of China replaced the Qing dynasty in 1911, the word 'China' became the formal abbreviation for the country.

The earliest state to appear in Chinese history was Xia. Its territory lay in the west of present day Henan province and in the south of present day Shanxi province. The remaining areas of the middle and lower reaches of the Yellow River were still inhabited by primitive societies of tribes or tribal leagues during the period of the Xia. At the time of the Shang and Western Zhou dynasties the territory under dynastic control was extended to the middle and lower reaches of the Yellow River, and to some areas in the middle and lower reaches of the Yangtze River.

When the first Emperor of the Qin dynasty unified China he established a vast country, extending from the sea in the east to the Qinghai and Gansu Plateau in the west, and from the Nan Ling Mountain Range in Guangdong in the south to the Hetao (the top part of the Great Bend of the Yellow River), Yinshan, and Liaodong in the north. In order to effectively rule such an extensive area, the Qin divided the whole country into 36 prefectures, each of which contained many counties. All of the prefectures were under the direct jurisdiction of the central government. Thus a system of prefectures with the county as the local administrative unit was set up. Although the names of the prefectures and their boundaries have varied with different dynasties, the county has remained as the basic local administrative unit ever since. There were only several hundred counties in the Qin dynasty, but the total reached about 1500 in the Han dynasty. Following the end of the Han dynasty the number of counties fluctuated around 1200 and then finally increased to over 2000 during the time of the Republic of China.

The province as an administrative unit, as seen in the current system, originated in the Yuan dynasty. Apart from Tibet and the North-East, there were 11 provinces in the Yuan dynasty. This system was carried on in the Ming and Qing dynasties, but the number of provinces was increased to 18 at the beginning of the Qing dynasty, and to 23 at the end of that dynasty, excluding Tibet and Inner Mongolia. Since the founding of the People's Republic of China, the province and the county have remained unchanged as the two main levels of administrative division. At the county level adjustments have been made annually according to the prevailing condi-tions, while at the provincial level there has been relative stability since the early 1950s, when considerable adjustments were made. There are now 22 provinces (including Taiwan), five minority autonomous regions, and three municipalities of provincial level in China (see Fig. 2.1 and Table 2.2 for the administrative division of China at the end of 1987).

The earliest historical record of population and land area can be found in the period of the Western Han dynasty. This record shows that in the year AD 2 there were 59 590 000 people registered, and the cultivated land amounted to 576 450 000 mu (38 622 150 ha). This gives a figure of 9.67 mu or 0.65 ha of cultivated land per person. In AD 755, during the prosperous times of the powerful Tang dynasty, the total registered population was 52 910 000 and there was an average of 27.03 mu (1.81 ha) of cultivated land per person. These figures included the population and land of conquered areas, but also reflected the depletion of the population by war over the centuries. In 1393, during the period of the Ming dynasty, the population had increased to a total of 60 540 000 and there was an average of 14.05 mu (0.94 ha) of cultivated land per person. The Chinese population first exceeded 100 million during the early stage of the Qing dynasty, and in 1708 it was

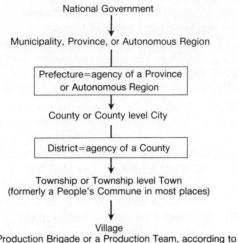

FIG. 2.1   The political/administrative levels in China, 1987.

recorded at 103 780 000. The average area of cultivated land per person at this time had declined to 5.50 mu (0.37 ha). Half a century later, in 1766, the population had increased to 208 090 000 and the average area of cultivated land per person had fallen to 3.70 mu (0.25 ha). In 1841, during the First Opium War, the total Chinese population was recorded at 413 450 000 and the area of cultivated land at 1.65 mu (0.11 ha) per person.

The above statistics were calculated on the basis of the number of taxpayers within the territory ruled by the respective governments at different times. They may not be particularly accurate, but they ought to be close to the true figures and should give a realistic indication of the size of the population and the area of cultivated land available over the centuries. They show clearly that increases in population did not parallel increases in the amount of cultivated land, even in ancient China, and this discrepancy grew more serious as time went on. As a result of this trend, an increasing population existing on less cultivated land became a distinctive characteristic of Chinese agriculture.

According to the statistics published by the State Statistical Bureau, the total population of China in 1949 was 541 670 000; the cultivated land totalled 1 468 220 000 mu (98 370 740 ha); and the cultivated land per caput was 2.71 mu (0.18 ha). By 1982, the population had increased to 1 015 410 000, yet the cultivated land had been extended by only 10 880 000 mu (728 960 ha) to a total of 1 479 100 000 mu (99 099 700 ha), and the cultivated land per caput had decreased to 1.45 mu (0.10 ha) (Tables 2.3 and 2.5).

In recent years the rapid population increase has been brought more under

TABLE 2.2 The administrative divisions of China, 31 Dec. 1987

	Prefectures	Cities			Counties	Districts under city administrations
		Total	Prefecture level	County level		
Municipalities						
Beijing					8	10
Shanghai					10	12
Tianjin					5	13
Provinces						
Anhui	7	16	9	7	65	35
Fujian	3	11	6	5	59	17
Gansu	9	13	5	8	67	10
Guangdong	4	18	11	7	92	27
Guizhou	7	8	2	6	75	6
Hebei	9	18	9	9	131	36
Heilongjiang	4	19	10	9	60	63
Henan	5	18	12	6	111	40
Hubei	7	25	8	17	54	27
Hunan	7	24	6	18	80	24
Jiangsu		21	11	10	54	42
Jiangxi	5	12	6	6	77	17
Jilin	2	14	6	8	33	18
Liaoning		19	13	6	39	55
Qinghai	7	2	1	1	37	4
Shaanxi	6	8	4	4	89	14
Shandong	5	25	10	15	86	33
Shanxi	6	10	5	5	96	17
Sichuan	9	19	11	8	174	31
Yunnan	15	11	2	9	114	4
Zhejiang	2	18	9	9	58	22
Autonomous Regions						
Guangxi Zhuang	8	11	5	6	77	21
Inner Mongolia	8	16	4	12	72	16
Ningxia Hui	2	4	2	2	16	6
Tibet	6	2	1	1	76	1
Xinjiang Uygur	13	16	2	14	71	11
Total	156	378	170	208	1986	632

Source: State Statistical Bureau (1988). *China Statistical Yearbook 1988*. International Centre for the Advancement of Science and Technology Ltd, Hong Kong, and China Statistical Information and Consultancy Service Centre, Beijing. (In English)

TABLE 2.3 The population of China (at year end)

Year	Total population (10⁶)	Urban population (10⁶)	Rural population (10⁶)	Urban population as a percentage of total population (%)
1949	541.67	57.65	484.02	10.6
1952	574.82	71.63	503.19	12.5
1957	646.53	99.49	547.04	15.4
1962	672.95	116.59	556.36	17.3
1965	725.38	130.45	594.93	18.0
1970	829.92	144.24	685.68	17.4
1975	924.20	160.30	763.90	17.3
1978	962.59	172.45	790.14	17.9
1980	987.05	191.40	795.65	19.4
1981	1000.72	201.71	799.01	20.2
1982	1015.41	211.54	803.87	20.8
1983	1024.95	241.26	783.69	23.5
1984	1034.75	330.06	704.69	31.9
1985	1045.30	382.44	662.86	36.6
1986	1053.97	434.29	619.68	41.2
1987	1069.16	497.77	571.39	46.6

Note: The total population figure for 1986 appears to exclude 3 240 000 servicemen. The urban population is defined as the population living in areas under the jurisdiction of cities and towns. In 1983 new regulations made it easier for peasants to move to market towns to work, and the sharp increase in the urban population from 1984 onwards is attributable mainly to the establishment of many new towns after adjustments in the criteria for defining towns (see Appendix 3). It should be noted that many people engaged in agriculture live and work within the designated boundaries of the cities and towns and make an important contribution towards the food supply of their cities and towns.
Source: State Statistical Bureau (1985). *Statistics of Chinese Society*. Chinese Statistical Publishing House, Beijing; State Statistical Bureau (1988). *China Population Statitics Yearbook 1988*. China Outlook Press, Beijing. (In Chinese)

control (Table 2.4). However, the distribution of population, especially between urban and rural areas, needs adjusting. At least 30 million people are reported to have taken part in inter-provincial migration since the founding of the People's Republic of China. The migration organized by the government mainly involved movement of technicians and skilled workers from cities, industrial bases, and mines in the eastern part of China to the inland or border provinces which needed economic reconstruction. This organized migration added new economic vitality to relatively backward areas. Spontaneous migration also took place, mainly from densely populated rural areas, such as the North China Plain in Henan and Hebei

TABLE 2.4   The annual growth of the population of China

Years	Total population (10⁶)	(%)	Urban population (10⁶)	(%)	Rural population (10⁶)	(%)
1950–52	11.05	2.0	4.66	7.5	6.39	1.3
1953–57	14.34	2.4	5.57	6.8	8.77	1.7
1958–62	5.28	0.8	3.42	3.2	1.86	0.3
1963–65	17.48	2.5	4.68	2.5	12.80	2.3
1966–70	20.91	2.7	2.76	2.0	18.15	2.9
1971–75	18.86	2.2	3.21	2.2	15.65	2.2
1978	12.85	1.4	5.76	3.5	7.09	0.9
1980	11.63	1.2	6.45	3.5	5.18	0.7
1981	13.67	1.4	10.31	5.4	3.36	0.4
1982	14.69	1.5	9.83	4.9	4.86	0.6
1983	9.54	0.9	29.72	14.0	− 20.18	− 2.5
1984	9.80	1.0	88.80	36.8	− 79.00	− 10.1
1985	10.57	1.0	52.38	15.9	− 41.83	− 5.9
1986	8.67	0.8	51.85	13.6	− 43.18	− 6.5
1987	15.19	1.4	63.48	14.6	− 48.29	− 7.8

Note: Some sources indicate an increase of 11 910 000 in the total population in 1986 (servicemen appear to have been left out of the 1986 total in the references quoted here). See note to Table 2.3.
Source: State Statistical Bureau (1985). *Statistics of Chinese Society*. Chinese Statistical Publishing House, Beijing; State Statistical Bureau (1988). *China Population Statitics Yearbook 1988*. China Outlook Press, Beijing. (In Chinese)

provinces, and Sichuan and Zhejiang provinces, to provinces where land resources were more readily available, or where more people were needed to develop industry, mining, and forestry, such as Heilongjiang, Inner Mongolia, Qinghai, Xinjiang and even Yunnan and Guangxi. Voluntary immigrants to Xinjiang from other provinces made up about two-thirds of the total immigrants to Xinjiang. Thus the average population density of the whole country increased by 72 per cent in the period 1949–79, but that of Heilongjiang, Inner Mongolia, and Xinjiang all increased by more than 150 per cent. This modified a little the general pattern of population distribution, in which the population in the eastern part of China is far denser than that in the west, and the population in the south is denser than that in the north. The tendency for the population to become concentrated in the cities, especially the big cities, has not been halted effectively, although the government has taken many measures to restrict this type of movement.

In recent years, the respective contributions of agriculture and industry to the national economy have changed substantially, and these changes have created an urgent need to transfer a part of the agricultural labour force out of cropping into other industries. With the policy of 'leave the land, but stay

TABLE 2.5   The cultivated land of China

Year	Total cultivated land (10^8 mu)	Paddy field (10^8 mu)	Cultivated land other than paddy (10^8 mu)	Irrigated land other than paddy (10^8 mu)
1949	14.68	3.42	11.26	0.48
1952	16.19	3.88	12.31	0.73
1957	16.77	4.13	12.64	1.60
1962	15.44	3.58	11.86	1.29
1965	15.51	3.71	11.80	1.52
1970	15.17	3.78	11.39	2.18
1975	14.96	3.83	11.13	3.10
1978	14.91	3.81	11.10	3.38
1979	14.92	3.81	11.11	3.34
1980	14.90	3.80	11.10	3.36
1981	14.86	3.79	11.07	3.34
1982	14.79	3.77	11.02	3.34
1983	14.75	3.76	10.99	3.25
1984	14.53			
1985	14.53			
1986	14.43	3.76		
1987	14.38	3.76	10.62	3.18

1 mu = 0.0667 ha or 0.1647 acres
Source: Society of Technological Economics of China (1986). *Handbook of Technological Economics*. Liaoning People's Publishing House, Shenyang; State Statistical Bureau (1987). *Rural Statistical Yearbook of China 1987*. Chinese Statistical Publishing House, Beijing; State Statistical Bureau (1988). *Rural Statistical Yearbook of China 1988*. Chinese Statistical Publishing House, Beijing. (In Chinese)

in the village' the rural labour force employed in non-agricultural activities increased by 45 million people during 1978–86. But there are at least 100 million more people whose labour is surplus to the needs of agriculture and who are waiting to make such a transfer; the problem of adjusting the distribution of the population to the developing economic situation still remains unsolved.

In 1949 agriculture made up 70 per cent of the gross output of agriculture and industry. By 1952 this figure had fallen to 56.9 per cent, though the agricultural (or primary industries) labour force constituted 83.5 per cent of the total labour force. Thirty-three years later, in 1985, agriculture contributed 34.3 per cent to the rapidly increasing gross output of agriculture and industry, and the agricultural labour force had fallen to 62.5 per cent of the total labour force. Despite this fall, the total number of people in the agricultural labour force in 1985 stood at 312 million, that is, 139 million more than in 1952, and there had been an 11 per cent decrease in the available area of cultivated land over that period. Where is the way out for this surplus agricultural labour force?

Chinese nationalities and customs

There are 56 nationalities in China, with the main nationality, the Han, constituting 93 per cent of the population. At the census in 1982, 15 minorities each had a population in excess of one million. They were the Zhuang, Hui, Uygur, Yi, Miao, Man, Tibetan, Mongol, Tujia, Buyi, Korean, Dong, Yao, Bai, and the Hani — listed in descending order of population. The Zhuang, who live mainly in Guangxi province, have a population of some 13 million, while the smaller minorities like the Olunchun and the Hezhe who live in the border areas of North-East China, and lived on hunting and fishing in the past, possess only several thousand people each (Table 2.6).

TABLE 2.6 Nationality constituents of the population of China, at 1 July in the census years 1953, 1964, and 1982

Nationality	1953 population		1964 population		1982 population	
	(10^6)	(%)	(10^6)	(%)	(10^6)	(%)
Nationalities with a population of more than one million in 1982:						
Han	545.28	93.92	651.30	94.23	936.70	93.30
Zhuang	6.61	1.14	8.39	1.21	13.38	1.33
Hui	3.56	0.61	4.47	0.65	7.22	0.72
Uygur	3.64	0.63	4.00	0.58	5.96	0.59
Yi	3.25	0.56	3.38	0.49	5.45	0.54
Miao	2.51	0.43	2.78	0.40	5.03	0.50
Manchu	2.42	0.42	2.70	0.39	4.30	0.43
Tibetan	2.78	0.48	2.50	0.36	3.87	0.39
Mongol	1.46	0.25	1.97	0.29	3.41	0.34
Tujia			0.52	0.08	2.83	0.28
Buyi	1.25	0.22	1.35	0.20	2.12	0.21
Korean	1.12	0.19	1.34	0.19	1.76	0.18
Dong	0.71	0.12	0.84	0.12	1.43	0.14
Yao	0.67	0.12	0.86	0.12	1.40	0.14
Bai	0.57	0.10	0.71	0.10	1.13	0.11
Hani	0.48	0.08	0.63	0.09	1.06	0.11
Other nationalities:						
	4.29	0.73	3.48	0.50	6.89	0.69
Total	580.60	100.00	691.22	100.00	1003.94	100.00

Source: State Statistical Bureau (1985). *Statistics of Chinese Society.* Chinese Statistical Publishing House, Beijing. (In Chinese)

Although the minority nationalities make up only a small proportion of the whole population, they occupy some 50–60 per cent of the total territory of China. They are spread mainly over the western part of China, but some live in the coastal islands or inner lands. The Han nationality can be found nearly all over the country, although most of them are concentrated in the eastern part of China. Because the population is always moving and diffusing, it is difficult to draw a distinctive line between two different nationalities: to say that a nationality lives in a certain area only means that people of a certain nationality are relatively concentrated there. There is always another nationality mingled in, and its population will not necessarily be less than that of the nationality whose name is given to that area. There are five Autonomous Regions of different minorities, at the provincial level, in present day China. They are: the Inner Mongolian Autonomous Region, established in 1947; the Xinjiang Uygur Autonomous Region, established in 1955; the Guangxi Zhuang Autonomous Region, established in 1958; the Ningxia Hui Autonomous Region, established in 1958; and the Tibetan Autonomous Region, established in 1965. There are also autonomous areas at lower administrative levels: in 1983 there were 590 autonomous counties or units equivalent to counties, and of these 264 were outside the Autonomous Regions mentioned above; of the 264, 73 were in Yunnan province where 23 minority nationalities live, and the remainder were scattered in various parts of China.

Their different historical backgrounds have made the nationalities unequal in their social and economic development. Their customs and religions also differ from one to the other. When the People's Republic of China was founded there were four million people in different parts of China, including the Tibetans in Tibet and the Tai in Yunnan, still living under a system of feudal serfdom. There were also about one million people, mainly of Yi nationality in Liangshan, Sichuan province, who were living in a slave type of society. Other minority nationalities were living in primitive tribal communities, these included the Olunchun who lived by hunting and fishing in the Da Hinggan Ling Mountains, Heilongjiang province (now, after boundary changes, in Inner Mongolia), and the Nu and the Wa who lived mainly by agriculture, in the Hengduan Shan Mountains of Yunnan province. The total population of these groups was about six hundred thousand. In contrast, the minority nationalities which had lived together with, or kept in contact with, the Han nationality over a long period had approached the same stage of social development as the Han, through mutual influence and cooperation.

The western part of China, including the plateaux, mountains, and basins of Inner Mongolia, Xinjiang, Qinghai, Gansu, Tibet, and north-western Sichuan, contains the main grazing land of China, and here the Mongolian, Tibetan, Kazak, and Tajik nationalities live. Animal husbandry is the main economic activity in this area and provides a livelihood for these minority peoples.

This pastoral area of China lies on the west side of the line which starts at the eastern foot of the Da Hinggan Ling range and passes through to the eastern edge of the Qinghai–Tibetan Plateau. This line does not clearly distinguish the grazing land from the cropping land; there is a transitory zone where the two activities overlap. Over the centuries this transitory zone has moved northward, and although it is still moving, the rate of movement is slowing down. This is because the rainfall is lower in the north, and desertification is now becoming serious, so that it is increasingly difficult to ensure profitable cropping at this margin. There are still some places in the arid north-west where the existing irrigation system can support a certain amount of cropping; on some of the oases in Xinjiang province cotton cultivation was begun about 1500 years ago and some other crops such as grapes and lucerne were first grown there more than 2000 years ago.

The minority nationalities have made a considerable contribution to the national economy, especially to the development of agriculture. For example, the donkey and the mule, commonly used in northern China, probably spread from the stock of the Hun nationality (Xiongnu). The Hun first tamed the donkey and domesticated it, then mated it with the horse to produce the mule. It is believed that during the Qin or Han dynasties the donkey and the mule spread into inland China where the Han nationality learned the skills of raising and breeding them. However, these animals remained rare in Yunnan and Guizhou provinces in South-West China down to the time of the Tang dynasty. It was not until the Yuan dynasty unified China that both the donkey and the mule were found all over China.

Kaoliang (sorghum) may have originally been cultivated by minority nationalities in South-West China, and barley is considered to have been first acclimatized by the aboriginals on the Qinghai–Tibetan Plateau. Highland barley, the staple crop of Tibet, has long been cultivated. It is generally agreed that China is the place of origin of the soya bean, but the exact site of its origin is still disputed. The biological behaviour of this crop, and the distribution of the wild varieties, suggest that the soya bean might have originated either in the border area of North-East China or in the border area of South-West China.

The patterns of agriculture or animal husbandry of the different nationalities dictated their lifestyle and also influenced their customs. The nomadic minorities moved about according to the availability of pasture and the climatic conditions, living in felt tents and obtaining food, mainly meat, cheese, and milk, from their herds and/or flocks. Clothes were made from sheepskins.

In contrast the Han nationality relied on cropping. Before the Han dynasty, their staple food was mainly millet, wheat, and barley in northern China, and rice in southern China. The grains from these crops were not crushed or ground, but were cooked whole to make 'fan'; that is why the Chinese always say, 'to have a fan' when they mean 'to have a meal'. Grinding wheat to make flour began in the Han dynasty, the practice is said to have been introduced from west Asia.

Tea, the most popular beverage in China, originated in the evergreen forest in the upper reaches of the Yangtze River, where people still collect leaves of the wild trees to make tea as before. The earliest record of drinking tea is found in the Chun Qiu period. Tea plantations were probably limited to Sichuan province at first, but as tea drinking grew in popularity tea was planted wherever it could be grown. By the time of the Tang dynasty, tea drinking was so popular that tea had become one of the daily requirements of the common family — along with firewood, rice, cooking oil, and salt.

China also has a long history of fermenting wines and distilling spirits, using various raw materials, yeasts, and production methods. These products full into three categories: naturally fermented fruit wines, pressed wines, and distilled spirits, and this is the order in which they appeared in historical times.

Drinking vessels have been found among Neolithic remains, and the ancient literature records that the nobility of the Shang kingdom liked drinking alcoholic beverages. A bronze wine pot, containing 3000-year-old wine made with yeast, was found in an ancient tomb unearthed in 1974 in Pinshan, in Hebei province. In wine making with yeast, sugar and starch fermentation can be combined together — in modern terminology this is called the compound fermentation method. This was quite an advanced method at the time of the Shang kingdom. Improvements in wine making led to a steady increase in the number of varieties of wine produced. Foreign experience was absorbed by the Chinese — grape wine was introduced into China from the West during the Han dynasty and the Chinese soon learnt to make grape wine themselves. The distilled spirit popular in present day China was introduced in the Yuan dynasty from Arabia.

The ancient clothing of the Han nationality can be deduced from the cultural relics unearthed in recent years. In the Western Zhou period the typical dress of the Hua-Xia nationality comprised a jacket called a 'yi', with the left front pulled to the right and a band fixed around the waist, and a skirt called a 'shang'. The hair was worn bundled up. By the Warring States period, the jacket and skirt were connected into a suit. After the Qin and Han dynasties the suit developed into a long gown with a large collar and loose sleeves. However, labourers at that time always wore short clothes for convenience. The nomadic people in the northern grazing lands dressed in a short coat, long trousers, and boots, which were convenient for riding and hunting. Later, the Han nationality adopted the dress of the nomadic people for their soldiers. Officers and women wore additional ornaments to indicate their high social status and show their good taste.

In early times most of the clothes were made of silk and hemp fibre; it was only after the Song and Yuan dynasties that cotton began to be the main raw material for fabrics. The silkworm was first bred in China, and silk fabrics can be traced back to the Shang dynasty. However, silk fabrics were expensive and were mainly for the nobility and rich men. Clothes worn by the common people were made of hemp or other plant fibres. In Chinese, fabrics made of plant fibre are called 'bu', while fabrics made of silk are called 'chou' or 'duan'. Since the common people used to wear clothes made of plant fibre the words 'bu yi' (cotton dress) became used for the common people.

Cotton did not originate in China. Although it was planted in Xinjiang some 1500 years ago, it was not introduced into the main eastern part of China until the end of the Song or the beginning of the Yuan dynasties. This probably occurred along two routes: from the south-east toward the middle and lower reaches of the Yangtze River, and from the north-west toward the Guanzhong Valley, in Shaanxi province. A woman labourer at the beginning of the Yuan dynasty contributed much to the improvement of spinning and weaving techniques in China. She had spent several years living on Hainan Dao Island, Guangdong province, and learned spinning and weaving from the Li nationality there. She then returned to her home town, improved on what she had learnt, and taught other people her methods. With her help the Sonjiang area (on the south side and adjacent to Shanghai) became famous for both its techniques of production and for the quantity of cotton cloth produced. The development of cotton cloth production in turn promoted cotton planting in China. A tradition was gradually established whereby the male members of a family would work in the fields cultivating the crops and the female members would stay at home spinning and weaving. This was an important characteristic of the self-sufficient economy of the Chinese rural areas in the past. The cloth thus woven was exported as far as the United States and Great Britain in the first 30 years of the nineteenth century: more than 1 000 000 bolts of cloth were exported through Guangzhou in some years. After the 1870s, however, cotton yarn and cloth spun and woven in Europe gradually replaced Chinese goods in Europe, and penetrated through the newly opened trading ports into inland China, presenting a powerful challenge to China's traditional production methods.

The historical achievements and limitations of traditional agriculture

Like Chinese culture, Chinese traditional agriculture had an early origin and progressed rather slowly; but it continued to progress over a period of several thousand years, and was sometimes in an advanced position compared with other countries. The system of cropping which gradually formed under the natural and social conditions of China was intensive and meticulous, featuring great attention to detail in such things as weeding and tilling the soil. Despite the relative shortage of land and surplus of labour which appeared in the eighteenth century, technical developments meant that agricultural production was still able to satisfy the needs of the increasing population. Indeed, the yields were approaching the maximum possible for traditional techniques. But this achievement was attained at the expense of an excessive labour input and damage to the environment.

Primitive agriculture appeared in China nearly 10 000 years ago, but the early stage was a transition period from collecting and hunting to cropping. Strictly speaking, the natural conditions of China are not very favourable for cropping. Tibet is too high and North-West China is too dry, and neither of these areas is suitable for cropping on a large scale; even the eastern part of China has its drawbacks. The annual rainfall in North China is not always sufficient, either in time or distribution, and spring drought appears in most years. In southern China there are too many mountains and hills, and swamps occupy a large part of the rest of the land; problems are also caused by the clayey soils and dense weeds. That is why dry farming techniques and production of drought-tolerant crops, such as millet and proso-millet (broom corn millet), are emphasized in North China, and rice planting is dominant in southern China. These natural conditions also explain the important roles of both irrigation and drainage in Chinese agriculture.

Three stages can be discerned in the development of Chinese agriculture: the experienced beginning stage, the developing stage, and the deepening stage. The Han nationality, who now live mainly by cropping, also engaged in animal husbandry 2000 years·ago. But as it became increasingly difficult to find sufficient grazing land, the proportion of animal husbandry became smaller, until it finally became a minor activity. Although the Chinese invented the plough at an early stage, it was initially drawn by man. There is archaeological evidence that the Chinese knew how to use cattle to draw ploughs as early as the Shang kingdom, but this did not become a common practice, and does not appear in the written record until the Chun Qiu and Warring States periods. In ancient times, cattle were used as a source of meat or as an offering on the sacrificial altar. Cattle were used as draught animals with the introduction of the iron plough, which was harder to pull than the wooden plough. However, draught cattle could only be used because man had learnt the key technique in controlling the ox — inserting a ring in the animal's nose. The importance of the nose ring in leading the ox was probably no less than that of a saddle in effectively controlling a riding horse, and it played a similar role in taming and training the ox.

Once the iron plough had been adopted the whole appearance of agriculture changed. The iron plough could be used to plough more land in a given time to a higher standard than the wooden plough, but more particularly, it provided the possibility for individual households to manage

FIG. 2.2 Ploughing using an ox controlled with a nose ring, Ming dynasty.
Source: Song Yingxing (1637). *Tian Gong Kai Wu*. From a facsimile of the first edition, in the Oriental Collections, British Library, Ref. 15226 b. 19.

their own cropping. Only by using animal power and an iron plough could an individual household carry out its field work to a high standard at the right season. In turn, as a basic unit of production, the individual household had greater adaptability and was more flexible than other forms of farming organization. That was why, over the years, individual farm households continued their activities obstinately, under various unfavourable conditions. The members of these households would work as hard as they could endure, only if they could cultivate land, and there was something left for their living. This type of agriculture was better than that carried out by the serfs of feudal lords, yet the scale of production was so small that the development of agriculture took the obvious course of increasing the yield per unit area of cropland. To do this a greater input of labour and more fertilizers were required, and the agricultural practices which developed under this regime constitute what may be called meticulous cultivation.

This system of agricultural practices existed in the Chun Qiu and the Warring States periods. Though not complete, or correct in many aspects, it did possess the following characteristics:

1. It was already known that farm work had to be carried out at the correct time in relation to the seasons.

2. Careful cultivation, or deep ploughing and crushing clods to smaller crumbs, was necessary to make a good seed bed.

3. Manures, such as human excrement and grasses, should be added to make the soil fertile.

4. Land use should be made more effective. Ridge culture was already being extended in some places. This practice of building ridges on which to sow the seed improves ventilation, facilitates sunlight reaching plant leaves, and improves the ability to prevent drought and waterlogging.

5. The cultivation system should be reformed. In places where there was a comparatively intensive cultivation system, the original cultivation system of leaving the land waste after several years' cropping had already been abandoned. Instead, different crops were planted in rotation, enabling more effective and efficient land use.

This was the beginning of Chinese traditional agricultural techniques featuring meticulous cultivation.

The succeeding period of some 1500 years, from the Qin dynasty to the Yuan dynasty, or from the third century BC to the beginning of the thirteenth century AD, was the developing stage of Chinese agriculture. During this period the regionalization of agriculture occurred, and the cropping techniques for the non-rice crops in North China, as well as for the rice crops in southern China, were systematized. The outstanding historian, Sima Qian, of the Western Han dynasty (first century BC) described accurately the general situation of three major agricultural regions. This was only possible after the unification of the First Empire of China in the Qin dynasty, and considerable progress had been made in agriculture. At that time in the areas around the middle and lower reaches of the Yellow River, where dry-crop cultivation was the main activity, industry and commerce had also developed in addition to the high level of agriculture. In southern China, or areas south of the Huai He River, where paddy rice cropping was the main activity, the land was not fully exploited and the techniques used were rather backward. A rice paddy was usually exploited by first clearing the trees and bushes, firing them, and then irrigating the area ready for the rice seed. After the rice seedlings had come up the field was flooded to drown the weeds. At that time southern China as a whole was in an early stage of development; population was sparse and there were no very rich or very poor men. The third main agricultural region was the expansive grassland area over China's north and west, where animal husbandry was the main activity, though in some places there was also some cropping. There were also transition zones between these regions.

The pattern of three main agricultural regions was thus formed early in the history of China, and has lasted until the present time without substantial change — though there have been modifications along the boundaries of these regions. The Qin and Han dynasties both established their capital in North China. They expanded the irrigated area around the capital, but they could not increase agricultural production over the whole of North China by such measures. Irrigation schemes were not as important to the early Chinese nation as some authors have claimed, at least not in the north. Through long years of practice the peasants accumulated a great deal of experience in dry-land cropping, and in the sixth century they finally formed a complete system of dry-land cropping embodying soil moisture conservation. The main feature of this system was that two additional activities designed to press the soil down after ploughing were included; the first was to crush the clods, and the second was to gently compact the soil. Although the people knew little about the principles of soil water movement, they did find that this method could decrease the evaporative loss of soil moisture. During the windy dry springs, which occur most years in North China, evaporative loss is normally so high that the soil moisture content becomes inadequate for seed germination or seedling growth. This cultivation system was therefore very important for North China since it helped the growth of seedlings.

In southern China there had also been great progress in the development of rice cropping techniques by the time of the Tang dynasty. Firstly, the straight-beamed plough was replaced by the curved-beamed plough. This new plough was more light and flexible, the depth of ploughing could be adjusted, and a mouldboard turned the soil upside down. This type of plough was introduced into Europe in the eighteenth century and has subsequently played an important part in improving the European plough. Other implements concerned with crushing clods and stirring mud were also invented, together with water-lifting and irrigation machines driven by man, animal, or water power.

During the Tang and Song dynasties some new measures were taken to alleviate the pressure of population on the land. One example was the reclamation of marsh land from riverside or lakeside depressions to form embankment areas after most of the surrounding arable land had been exploited. Embankments were constructed such that subsequent cultivation within the enclosed area could be carried out below the water level outside the embankments. In addition, hillsides were terraced, and earth dykes were constructed to control soil erosion and water loss. Without question these were all good, well designed, measures, but when they were extended excessively to unsuitable places in later times they proved to be harmful to water courses, and to soil and water conservation.

The achievements in agricultural technology of the Ming and Qing dynasties, the last two dynasties in Chinese feudal history, remain subjects of some dispute. Some scholars claim that this was a period of stagnation in agricultural technology because there was little improvement in either implements or practices. This may be correct in some respects, but it is not necessarily so when agriculture is taken as a whole within a broader context. The important point is that the population of the country increased rapidly and the contradiction between a growing population and limited arable land was becoming so serious that all the agricultural practices had to be used to solve this problem — and there were considerable achievements. The practices that were used are still of importance today, so they should be discussed in a little more detail.

In the areas around Hangzhou, in Zhejiang province, and Guangzhou, in Guangdong province, where the population was dense, the people began an integrated management system which involved cropping grain, planting mulberry trees to raise silkworms, and fish farming. The important features of this system were that the central part of a depression was dug out to form a pond for breeding fish; the earth dug out was piled around the pond to form a base for planting mulberry trees to raise silkworms; and the silkworm waste was used to feed the fish. As a rule, the pond occupied 60 per cent and the base 40 per cent of the total area. This system not only fully utilized the local natural resources, but also kept the natural relations between people, livestock, and plants in an ecological balance, making all the productive elements benefit each other. At that time people also began to remove the sludge from the pond and river to fertilize the land. This sludge, consisting of fertile surface soil carried into the pond or river by heavy rain, the fish wastes, and decomposed grasses, was good manure which increased the fertility of the land, and supplemented the eroded surface soil.

Another substantial achievement at that time was the development and spread of successive, intercropping and inlaid cropping systems which raised the cropping index. Intercropping is a system in which two or more crops which grow at about the same time are planted to form a composite population of crops. Inlaid cropping is a system in which two or more crops with fundamentally different, but overlapping, growing periods are grown in the same field. In this system a crop can be sown before the previously sown crop is harvested. These cropping systems not only utilize space fully, but they also utilize the solar energy effectively. These systems were first used in vegetable gardens, and were then introduced into field cropping. They brought an increase in the intensity of management, and provided a good means of easing the problem of population pressure on the land.

The cropping system in North China, in which three crops are planted in two years, was probably started in the Tang dynasty and became the basic system in the Ming and Qing dynasties. It was very popular in Hebei and Shandong provinces, and in Guanzhong, in Shaanxi province. The basic form of this system was to plant a crop of grain and harvest it in autumn every year, and in addition to plant a winter crop once every two years.

The system of harvesting two crops every year was first introduced in southern China in the Song dynasty. The planting of rice in summer, followed by wheat, legumes, or vegetables in winter, was still the predominant cropping system in this region during the Ming and Qing dynasties. However, because of the higher yield of rice compared to other crops, and the longer frost-free period in southern China, the double rice cropping system was developed and extended during these dynasties.

Further south, towards Fujian and Guangdong, where the climate is even warmer, and the area is essentially snow and frost free, a cropping system of three crops each year was developed. In this system a dry crop is planted between the two rice crops.

There was also introduction of high yielding crops and breeding of better varieties of crop plants. Crops such as maize, sweet potato, and peanut, all of which originated in America, were introduced into China after the middle of the Ming dynasty.

Maize was originally introduced into only a small area of China; however, this area increased rapidly after the founding of the Qing dynasty, and maize was planted not only in North China, and in the middle and lower reaches of the Yangtze River, but by the middle of the nineteenth century it was also widely planted in North-East China. In some poor areas it became the staple food.

The sweet potato was introduced from the Philippines by local officers of Fujian province at the end of the sixteenth century for the purpose of relieving people in the province during drought. By the end of the Ming and beginning of the Qing dynasties it was being planted in most parts of China.

The introduction of these high yielding crops not only changed the pattern of cropping, but also promoted the development of new cultivation techniques to suit the natural and social conditions of China. Crop breeding techniques were also improved.

As farming became more intensive, cultivation techniques were followed with ever increasing attention to detail, with a resulting high labour input: production costs in terms of labour and manure were ignored.

Although there was little progress in exploiting resources of manure, or in improving implements, during this period of the Ming and Qing dynasties, traditional agriculture was always improving the use of the manure and implements available. That the implements were not improved very much, did not necessarily mean that there was no progress at all.

For example, the traditional plough was not improved to till more deeply; instead, a new system of tillage was invented. The land was first ploughed using manual power, then ploughed again with animal power; alternatively a large plough pulled by a couple of oxen was used first, followed by a single plough and one ox. Another example is the application of manure. In addition to increasing the area manured, the 'three observations' system of applying manure was developed, which required the manure to be applied differently according to season, soil type, and crop. So that the limited amount of manure could have an optimum effect on crop growth, a dressing at the time the crop was growing was emphasized, in addition to the basic dressing. This second dressing was applied differently, according to the different conditions. On some occasions it was even applied separately for each individual plant.

It is obvious, then, that there were achievements in agricultural techniques, and that they related to the cropping system, the pattern of cropping, the cultural techniques, and the management methods. These achievements were a response to the problem of heavy population pressure on comparatively little land, and resulted in Chinese agriculture becoming more intensive.

The history of Chinese agriculture is very long, and its achievements are great. The agricultural techniques which developed over this long period of time, and which are still fundamentally being applied, have certain characteristics. They are not anachronistic, but still have relevance. However, they are not based on modern experimental science, and do not benefit from advanced industrialization. With these techniques alone, Chinese agriculture can not escape from the difficult position that it is in; for just as the achievements are substantial, the drawbacks are very apparent too. It is worth considering how these achievements and drawbacks are mutually dependent. Evaluating the past correctly can help toward a better understanding of present reality, and of the outlook for the future. In line with this thinking, let us then summarize the achievements and drawbacks, the positive and negative directions, of Chinese agriculture.

There are at least three particular aspects of Chinese agricultural achievements that are worth mentioning:

1. Successive cropping, intercropping, and inlaid cropping, multiplied the

cropping areas within the limited cultivated areas, and succeeded in feeding and in continuing to feed an ever growing population.

2. The practices of meticulous cultivation, manuring, and soil conservation, have ensured that soil fertility has been utilized effectively and maintained at a high level. Land is not exhausted after many years of cultivation.

3. The preparation and application of organic manures allows the wastes of the nation to be transformed into agricultural products capable of being utilized by human beings.

While the drawbacks are that:

1. Traditional agriculture over the centuries was mainly concerned with increasing the yield per unit area, or in other words with land productivity. Little attention was paid to the question of how to increase labour productivity. Under the labour-intensive system that developed, it was not considered whether the labour use was reasonable or not. The result of this system was that the requirement for food and the growth of population influenced each other, resulting in a vicious cycle. As the population pressure upon limited land increased, people tried to increase the yield, and the increase in yield was mainly dependent on an increase in labour input. As the labour force could only be increased by natural population growth, the pressure the population applied to the land became even heavier, and the conflict between more population and less cultivated land grew.

2. In traditional Chinese agriculture the inputs into agriculture are from within a closed cyclic system. This system can transform waste materials into useful products, but it limits the input of energy or power from outside the system. Mechanization of cultivation practices is not so urgent when there is an abundant labour force and a large population: mechanization can reduce the intensity of labour, but it cannot ensure an increase in production. On the other hand, agricultural practices carried out by manual power, meticulous though they may be, cannot cope with a serious natural challenge because the amount of power a human can supply is limited by physiological capability. This gives a low ability to control natural disasters, and the ecological environment may tend to deteriorate.

A large population — insufficient arable land — food shortage — that is the cruel reality which the Chinese people must face. Chinese agriculture has successfully survived the cycle of alternately rising and declining feudal dynasties, and conflicts among nationalities. Yet, it still suffers from the vicious cycles inherent in the technical systems employed, and in the disasters of the natural environment. How to find a way out remains a question to be considered seriously by everyone in China who wants to do something for his country.

References

The following texts are in English:

Bai Shouyi (1982). *An Outline History of China — From Ancient Times to 1919.* Foreign Languages Press, Beijing.

Barker, Randolph and Sinha Radha with Rose, Beth (Eds) (1982). *The Chinese Agricultural Economy.* West Special Studies in China and East Asia. Westview Press, Boulder CO.

Institute of the History of Natural Science (1983). *Ancient China's Technology and Science.* Foreign Languages Press, Beijing.

Bray, Francesca (1984). *Volume 6: Biology and Biological Technology, Part II: Agriculture.* In Joseph Needham (Ed.) *Science and Civilisation in China.* Cambridge University Press, Cambridge.

Perkins, Dwight H. (1969). *Agricultural Development in China 1368-1968.* Edinburgh University Press, Edinburgh.

State Statistical Bureau (1986). *China: A Statistical Survey in 1986.* New World Press, and China Statistical Information and Consultancy Service Centre, Beijing.

3

The components of agriculture

CHEN REN

With a section on The Fishery by

ZHANG LIN

Agriculture in a broad and narrow sense

The general meaning of agriculture in China is that agriculture consists of agriculture (i.e. cropping), forestry, animal husbandry, sideline activities, and fishery — that is, five types of activity. It should be noted that the term agriculture has two meanings in China: the broad meaning just described, or the narrow sense within that broad meaning as a synonym for cropping. The word agriculture in 'gross output of industries and agriculture', as quoted in the national statistics, denotes agriculture in the broad sense. The word agriculture in the name 'Ministry of Agriculture, Animal Husbandry and Fishery' (MAAHF), denotes agriculture in the narrow sense. (This name has recently been shortened to 'Ministry of Agriculture'.) To many Chinese, agriculture just means to cultivate land to grow crops. The people who grow crops may occasionally breed one or two pigs, keep a dozen or so fowl, plant trees around their house or village, or weave straw or bamboo articles in their spare time — but these are all minor activities. Of course, this is not the view of the minority nationalities who rely on animal husbandry. But they form only a very small proportion of the total population of China, and they make only a small contribution to the country's total production. These minorities, then, are an exception, and the Chinese as a whole usually look upon crop cultivation as the mainstay of agriculture.

Published government figures for the Gross Output Value of Agriculture (GOVA) since 1952, and the percentage contribution of the five components of agriculture are shown in Table 3.1. There was an increase in GOVA from 1952 to 1986, with the fastest relative increases being in the four components other than cropping, although the latter component also increased in real terms. In particular there was a noticeable increase in the percentage contribution of the sideline component from 1975 to 1984, caused by the rapid development of township enterprises. When this information was

73

TABLE 3.1   The Gross Output Value of Agriculture (GOVA) of China, and its components, 1952–86

Year	GOVA	Constituent components of GOVA				
		Cropping	Forestry	Animal husbandry	Sideline activities	Fishery
	(10⁹ yuan)	(%)	(%)	(%)	(%)	(%)
1952	41.70	83.1	0.7	11.5	4.4	0.3
1957	53.67	80.6	1.7	12.9	4.3	0.5
1965	58.96	75.8	2.0	14.0	6.5	1.7
1975	128.50	72.5	2.9	14.0	9.1	1.5
1979	158.43	66.9	2.8	14.0	15.1	1.2
1984	337.70	58.0	4.1	14.2	22.0	1.7
1986	401.30	62.2	5.0	21.8	6.9	4.1

Notes: The figures for 1952, 1957, and 1965 are based on constant prices of 1957; for 1975 and 1979 on constant prices of 1970; for 1984 on constant prices of 1980; and for 1986 on the prices of that year. The figures for 1986 exclude industrial production at township level and below which previously was included under sideline activities (see Chapter 6).

Source: Editorial Board of China Agriculture Yearbook (1986). *China Agriculture Yearbook 1985 (English Edition)*. Agricultural Publishing House, Beijing; Editorial Board of China Agriculture Yearbook (1988). *China Agriculture Yearbook 1987 (English Edition)*. Agricultural Publishing House, Beijing.

collected by MAAHF these township enterprises were not distinguished from the sideline activities; however, since 1984 the State Statistical Bureau, has excluded industrial production at village level and below village level from sideline production in the agricultural sector, and included it in the industrial sector figures. This change is reflected in the figures for 1986; in that year, the cropping component contributed just under two-thirds of the total production, while the contribution of the sideline component became quite small. The contributions of forestry and fishery to the total were always very small, though gradually increasing, and reached contributions of five per cent and four per cent respectively in 1986, with the calculations based on the new method. The contribution of animal husbandry was much larger than that of forestry or fishery, but still only about one-third that of cropping in 1986. This is quite different from the situation in many western countries, where the respective contributions of the cropping component and the animal husbandry component are about the same, or the latter is even greater than the former.

Within the scope of the cropping component, 'grain' production predominates.[1] The area sown (or planted) to 'grain' crops since 1952, as a percentage of the total sown area, shown in Table 3.2, has been decreasing slowly, but it still amounted to more than 75 per cent of the total sown area of the whole country in the mid-1980s. In the past, less than 10 per cent of the total sown area was sown to industrial crops, but in recent years this proportion has increased, and in 1985 and 1986 the figures were 15.5 per cent and 14.1 per cent respectively. The area sown to other crops, such as vegetables, fruit, and green manure, is quite small: it was about 5 per cent of the total sown area during the 1950s, increased slightly after the 1960s, and in the second half of the 1980s was about 7–9 per cent.

'Grain' production is, therefore, of special importance in Chinese agriculture. There are many reasons for this. The officially published figure for the total area of cultivated land is around one hundred million hectares, and the area of cultivated land per person, based on this figure, is only about 0.1 ha. Satellite information suggests that more than one hundred million hectares are under cultivation, but the official figure has not yet been amended. Even if the error is large, the fact remains that the area of cultivated land per person in China is still very small.

During the long course of historical development, nearly all of the potential arable land was cultivated. To expand the cultivated area further is now very difficult, and is no longer possible on a large scale. The present cultivated areas are scattered all over the country: many are situated in hilly or mountainous areas, artificially terraced right to the peaks. Much of this land is in small parcels, especially in the mountainous areas. The cultivated land is, in fact, distributed wherever peasants have been able to cultivate and thus acquire sufficient to live on, and under these circumstances it is understandable that the Chinese attach a special importance to 'grain' production. This attitude has been reinforced during the last 100 years by long wars and numerous disasters, which have often made it very difficult for common people to survive. With little industrial activity, they have had to live on the produce of these small amounts of land.

Since the 1950s there have been some changes. As mentioned above, the area sown to 'grain' crops has decreased by about 10 per cent, and the contribution of the cropping component to the GOVA has also fallen. This has resulted from the successful development of China's economy as a whole. With the further development of industries and other economic sectors, it is possible that the contribution of 'grain' production to the GOVA will continue to fall, but the fundamental fact that 'grain' produc-

1 In English language publications from China the word grain is used to include crops such as dry beans, potatoes, and sweet potatoes in addition to the normal cereal crops. This wider term 'grain' is printed here within single quotation marks to emphasize this difference of meaning. A more accurate term would be staple food crops — see next section.

tion occupies the dominant position in Chinese agriculture is not likely to be changed in the foreseeable future.

In the past, rural sideline activities comprised handicrafts, such as weaving wicker baskets; gathering wild plants; hunting wild animals; some forms of manual labour, such as bricklaying and carpentry; and some modern industrial activities and service businesses. Strictly speaking some of these sideline activities are not within the scope of agriculture, and there have been various changes in the definition of the term sideline activities. Since 1984 it has been much narrower in scope than previously. These activities will be dealt with further in Chapter 6.

Forestry is part of agriculture in its broad sense. In China it is administered separately by the Ministry of Forestry, and it will be discussed briefly again in the latter part of this chapter.

Animal husbandry and fishery also fall within the broad sense of agriculture, and they, too, will be dealt with separately in the latter part of this chapter. However, as they are of much less importance compared with crop cultivation they will only be dealt with briefly.

The classification of crops

There are various classification systems for crops. The one used most widely in China classifies all crops into three categories: 'grain' crops, industrial crops, and other crops.

Rice, wheat, and maize are the three main 'grain' (or staple food) crops, and in 1986 they occupied 73 per cent of the area sown to 'grain' crops. The next most important 'grain' crops are barley, millet, sorghum, beans, and sweet potato.

Chinese custom divides food into two categories — the main food, and auxiliary food. 'Main food' is that which is essential for human beings, and it generally denotes grain (i.e. rice, wheat, or maize). 'Auxiliary food' means that which helps make a good meal, but is not essential. Vegetables, meat, eggs, milk, and bean curd are all auxiliary foods. However, the State Statistical Bureau of China stipulates that bean seeds which are harvested after they have ripened and intended for human consumption, together with sweet potato, and potato, are all within the scope of Chinese 'grain'. In arriving at this statistic sweet potato and potato were converted in the ratio of 4:1 up to the end of 1963; since 1964 the ratio has been 5:1. That is, 5 kg of fresh sweet potatoes or potatoes are treated as the equivalent of one kilogram of other 'grain'.

Table 3.2 shows that the sown (or planted) areas of rice, wheat, and maize as percentages of the total sown area increased by 2 per cent, 3 per cent, and 4 per cent, respectively, during the period 1952–86. This is encouraging, since it implies that the infrastructure related to agricultural production has been

TABLE 3.2 The total sown area, and the proportion of the sown area planted to 'grain' crops, 1952–86

Year	Total sown area (10⁶ ha)	Proportion of the total sown area planted to 'grain' crops (%)	Proportion of the total sown area planted to particular 'grain' crops					
			Rice (%)	Wheat (%)	Maize (%)	Soya bean (%)	Potato and sweet potato (%)	Other 'grain' crops (%)
1952	141.3	87.8	20.1	17.5	8.9	8.3	6.2	26.8
1957	157.2	85.0	20.5	17.5	9.5	8.1	6.7	22.7
1965	143.3	83.5	20.8	17.2	10.9	6.0	7.8	20.8
1978	150.1	80.3	22.9	19.4	13.3	4.8	7.9	12.0
1983	144.0	79.2	23.0	20.2	13.1	5.3	6.5	11.1
1984	144.2	78.3	23.0	20.5	12.9	5.1	6.2	10.6
1985	143.6	75.8	22.3	20.3	12.3	5.4	6.0	9.5
1986	144.2	76.9	22.4	20.5	13.3	5.8	6.0	8.9

Note: The above figures include the multiple cropped areas.

Source: State Statistical Bureau (1986). *China: A Statistical Survey in 1986*. New World Press, and China Statistical Information and Consultancy Service Centre, Beijing (In English); Editorial Board of China Agriculture Yearbook (1986). *China Agriculture Yearbook 1985 (English Edition)*. Agricultural Publishing House, Beijing; Editorial Board of China Agriculture Yearbook (1987). *China Agriculture Yearbook 1986 (English Edition)*. Agricultural Publishing House, Beijing; Editorial Board of China Agriculture Yearbook (1988). *China Agriculture Yearbook 1987 (English Edition)*. Agricultural Publishing House, Beijing.

improved, and that the peasants have paid for increased inputs. The increase in the production of these crops also reflects an improvement in the living conditions of the rural people. However, the decrease in the percentage area sown to soya bean is not good, for soya bean is still one of the most important protein sources in the daily diet for rural people.

The second main category of crops is the industrial crops, sometimes also called cash crops or economic crops. The most important crop in this category is cotton. Next in importance are the oil-bearing crops, including peanut, rape, and sesame. After the oil crops come the fibre crops, which include jute, kenaf, hemp, and flax; and then follow tobacco and tea. The percentages of the total sown area devoted to these crops are shown in Table 3.3. The actual area sown to industrial crops increased from 12.56×10^6 ha in 1952 to 20.29×10^6 ha in 1986, with the area sown or planted to sugar crops expanding the most rapidly.

The third main category of crops is 'other crops'. The main crops in this category are the green manure crops, vegetables, and fruit. However, for statistical purposes, this category includes all crops that do not fall within the previous two categories.

Crops can also be classified according to their sowing season or harvesting

TABLE 3.3   The proportion of the total sown area planted to industrial crops, 1952–86

Year	Proportion of the total sown area planted to industrial crops (%)	Proportion of the total sown area planted to particular industrial crops					
		Cotton (%)	Oil crops (%)	Fibre crops (%)	Sugar crops (%)	Tobacco (%)	Other industrial crops (%)
1952	8.8	3.9	4.0	0.4	0.1	0.3	0.1
1957	9.2	3.7	4.4	0.4	0.3	0.3	0.1
1962	6.3	2.5	3.0	0.3	0.2	0.2	0.1
1978	9.6	3.2	4.1	0.5	0.6	0.5	0.7
1983	12.3	4.2	5.8	0.3	0.8	0.5	0.7
1984	13.4	4.8	6.0	0.3	0.9	0.6	0.8
1985	15.5	3.6	8.2	0.9	1.1	0.7	1.0
1986	14.1	3.0	7.9	0.5	1.0	0.8	0.9

Note: The above figures include the multiple cropped areas.
Source: Editorial Board of China Agriculture Yearbook (1986). *China Agriculture Yearbook 1985 (English Edition)*. Agricultural Publishing House, Beijing; Editorial Board of China Agriculture Yearbook (1987). *China Agriculture Yearbook 1986 (English Edition)*. Agricultural Publishing House, Beijing; Editorial Board of China Agriculture Yearbook (1988). *China Agriculture Yearbook 1987 (English Edition)*. Agricultural Publishing House, Beijing.

season. Classification by sowing season divides the crops into three categories. The first category, autumn-sown crops, includes winter wheat, winter barley, and winter rape. The second category is spring-sown crops, and includes cotton, spring sorghum, spring maize, and spring wheat. The third category of summer-sown crops includes summer maize, summer soya bean, and summer sweet potato.

Crops classified by harvesting season fall into two categories: summer-harvested crops and autumn-harvested crops. Wheat, rape, and barley are examples of summer-harvested crops; and maize, cotton, millet, and sorghum are autumn-harvested crops. Early rice is a summer-harvested crop, but middle and late rice are autumn-harvested crops. The State Statistical Bureau separates the summer-harvested 'grain' crops and the autumn-harvested 'grain' crops when publishing the 'grain' production figures. Both of these systems of crop classification are common in China and may readily be found in articles published in newspapers and magazines.

Multiple cropping systems

Since the sowing and harvesting seasons of crops differ, it is possible to plant crops with different sowing or harvesting seasons in such a way as to obtain more than one harvest a year from one piece of land. Chinese peasants have practised such multiple cropping for several hundred years, and are masters in this field: about half of China's cultivated land is now under multiple cropping. If expressed in terms of the sown (or planted) area, then two-thirds of China's crops are grown in this way.

There are three types of multiple cropping practised in China: intercropping, inlaid cropping, and successive multiple cropping.

Intercropping is a cropping system in which two or more crops with similar growing seasons are planted in a piece of land alternately according to a certain number of rows. For example, one of the most widely adopted inter-cropping systems in China is the intercropping of maize with soya bean, or of maize with mung bean. Much of the maize in China is grown under this system, and of the bean crop too. The specific pattern can be one row of maize alternated with one row of soya bean (or other bean), two rows of maize and one row of soya bean, or in some places, two rows of maize and two rows of soya bean, and other patterns may also be encountered.

Inlaid cropping is sometimes called relay intercropping. It is a system in which two or more crops with different growing seasons are planted in a piece of land in such a way that the second crop is planted before the first crop is harvested. That is, the growing seasons of the crops overlap, but they are not the same, as they are in the intercropping system. The inlaid cropping system is quite popular in China. In North China the most common inlaid cropping practised is maize inlaid-cropped into wheat. In the Yangtze River

Valley it is common for cotton to be inlaid-cropped into rape (or broad bean, or wheat); also maize is inlaid-cropped into wheat, and then sweet potato or soya bean is inlaid-cropped into the maize. In South China it is easy to find sugar cane inlaid-cropped into rice.

Successive multiple cropping (also called sequential cropping) is the system in which two or more crops are planted successively in the same piece of land in the same year — it has to be in the same year, otherwise it is just a form of single cropping. This system differs from inlaid cropping in that the second crop is planted after the first crop is harvested. Successive multiple cropping is the main form of multiple cropping in China, especially in areas south of the Yellow River. There are many different forms of successive multiple cropping. Three crops every two years is very popular in North China. The common form of this system is to plant a summer crop such as summer maize, summer soya bean, or summer sweet potato in the wheat field after the wheat harvest, then to leave the field fallow in winter, and to plant a spring crop such as spring maize, spring sorghum, or spring sweet potato in the spring. This crop is then followed by winter wheat sown in the autumn after the harvest of the spring crop. Thus, in two years, one crop of wheat, one summer crop, and one spring crop are harvested.

In places with a warmer climate, or more favourable conditions, the system of three crops in every two years is usually replaced with a system of two crops in one year. In most cases, a summer crop is planted after the wheat harvest, and wheat is sown again after the harvest of the summer crop in the autumn. There are also other successive multiple cropping systems practised, such as five crops in three years, and three crops in one year, particularly in southern China. The common form of the latter system is to plant two crops of rice successively and then a crop of winter wheat, or rape, or green manure.

The purpose of multiple cropping is to increase the harvest of products from a given area of cultivated land. This purpose is realized in two different ways. The first is through increasing the extent of time utilized; and the second is through increasing the extent of space utilized. Successive multiple cropping increases the utilization of time, intercropping increases the utilization of space, and inlaid cropping mainly increases the utilization of time but also increases the utilization of space.

Certain conditions have to be met in order to increase the utilization of time. First of all, the system has to be practicable under the local climatic conditions. In places where the winter is so cold that no crop can survive during the winter months, it is not possible to have successive multiple cropping. Providing the winter is not as cold as this, then the number of crops that can be harvested from a piece of land in a year will increase with the increasing length of the frost-free period. This is why only one crop per year is possible in the northernmost part of China; three crops every two years is a very popular system in the northern part of North China, though a small proportion of the land is under two crops a year; and two crops a year is the most common system in the southern part of North China. It is also why the dryland, south of the Yangtze River, is planted to two crops a year, while the rice fields there can be planted to three crops a year; and the system of three crops a year is common in South China. Secondly, a higher level of fertility of the soil is required, and/or higher applications of fertilizer are necessary, if reasonable yields are to be obtained without exhausting the land. Thirdly, more labour is required for multiple cropping, so more labour resources have to be available. Which kind of multiple cropping system is appropriate for a particular place, and how the tillage and cultivation practices are to be arranged, are questions to be determined according to the local conditions and the requirements of the people concerned.

Increasing the utilization of space is dependent on the biological behaviour of the crops. For example, maize grows high, while soya bean grows low. Planting one row or two rows of maize alternately with one row or two rows of soya bean can improve aeration as well as the incidence of solar radiation with certain plant densities, besides increasing the nitrogen supply via nitrogen fixation in the root nodules on the soya bean. The point here is that the maize utilizes the space which otherwise will not be utilized by the soya bean. Planting a row of maize every five meters or so in a sweet potato field, will provide an additional yield of maize without affecting the growth of the sweet potato. In fact, it is a general rule that crops of different behaviours should be planted in an intercropping system otherwise a benefit will not be produced.

Cropping index and crop yield

The utilization of time, in cropping land, can be expressed in terms of the cropping index. This indicates the number of times a crop harvest is taken from a piece of land in a 12-month period. The index does not take account of the number of different kinds of crop harvested: two crops grown together and harvested at the same time count as only one harvest. The cropping index is expressed as a percentage and is calculated as follows:

$$ \text{CI} = \sum_{i=1}^{n} N_i A_i \bigg/ \sum_{i=1}^{n} A_i \times 100 \text{ per cent} $$

where CI = cropping index

N_i = the average number of times per year there is a harvest from the piece of land i

A_i = the area of the piece of land i

n = the total number of pieces of land

If there is only one piece of land, then n = 1 and CI = $N_i \times 100$ per cent; one harvest a year gives a cropping index of 100 per cent, two harvests a year give a cropping index of 200 per cent. Note that N_i is the average number of harvests per year for a particular piece of land; where the system of three crops in every two years is practised there will be one harvest a year in one year and two harvests a year in the next, giving an average value of 1.5 for N_i.

The concept of the cropping index has obviously not been developed for a single piece of land. Within a given area or region different pieces of land will often be cropped according to different systems, even if one system may prevail. For example, although the system of three crops every two years prevails in North China, not every piece of land will be cropped under this system — some pieces of land may be used to grow just a single cotton crop a year, and so on. By using the cropping index concept a more precise, quantitative method of assessing time utilization in cropping is achieved.

Sometimes the term 'cropping index' is also used for the figure obtained by dividing the total sown area by the total cultivated area, for a certain year and defined region. If the region is large enough the results of both methods of calculating the cropping index are likely to be similar.

As explained above, intercropping does not increase the cropping index as the crops are harvested at more or less the same time. In calculating the yield from crops grown in this way, the yield of each crop is related to the proportion of the piece of land occupied by the particular crop. For example, if two rows of maize and one row of soya bean are intercropped, then the yield of the maize will be related to two-thirds of the field area, and the yield of the soya bean to one-third. From this it will be seen that the calculated yield per unit area of a crop grown in an intercropping system will usually be greater than if the crop was grown alone.

Inlaid cropping and successive multiple cropping do increase the cropping index as there will be more than one harvest time. But in calculating the yield from crops grown in this way it is assumed that each crop has been grown on the whole area of the field, and therefore the calculated yield is likely to be less for a particular crop than if that crop had been grown alone. For example, if maize is inlaid cropped into wheat in a field of area a, then the yield per unit area of wheat will be calculated from the total yield of wheat divided by a, and the yield per unit area of maize will be the total yield of maize divided by a. If the maize is planted after the wheat is harvested then it is obvious that the yields will be calculated separately.

To summarize, intercropping, successive multiple cropping, and inlaid cropping are all systems designed to increase total production, but because of the methods adopted for calculating agricultural productivity, any increase achieved will be represented differently between the systems. Intercropping will show an increase in yield per unit area for the individual crops, but will not show an increase in cropping index; inlaid cropping and successive multiple cropping will show an increase in cropping index, but not in yield per unit area for each individual crop.

If multiple cropping is to be adopted, be it intercropping, inlaid cropping, or successive multiple cropping, more inputs of labour and materials will be required, and therefore any proposal to begin multiple cropping has to be considered carefully. The intended multiple cropping system must be feasible under the local natural conditions, the additional requirements of labour and materials have to be available, and the new system must be economically beneficial. If these conditions are not met it is quite possible that the result will not be an increase in production even though more inputs of labour and materials have been utilised, or the increase in production may not be of economic benefit. In these situations it is not reasonable to introduce multiple cropping.

Multiple cropping is an important aspect of Chinese agriculture; and Chinese peasants have been willing to trade an abundance of labour for an increase in 'grain' production at the expense of a decrease in labour productivity — a situation quite different from that found in many Western countries. Even so, whether a specific multiple cropping system is reasonable or not has to be considered comprehensively, and over the years Chinese peasants have had both positive and negative experiences in this regard.

The main crops

Rice

China has the longest history of rice cultivation in the world, and the country includes one of the regions of origin of cultivated rice. Archaeological remains discovered in 1973 at Hemudu Village, Yuyao County, Zhejiang province, have been dated to 7000 BP, and indicate that rice was being cultivated at that time.

The present planted area of rice is about 33×10^6 ha over the whole country, and this yields about 170×10^6 tonnes. (In this context, if 1 ha of land is planted with two crops of rice in one year, the planted (or sown) area of rice is 2 ha.) Rice is not only the largest crop grown in China, but the Chinese rice crop occupies an important position in world rice production. The planted area of rice in China is second only to that of India, and the total production of rice far surpasses that of India, and accounts for more than one-third of total world production.

Rice is a hydrophilic and thermophilic crop. Its minimum temperature for germination is 10–12°C, and for earing 20–25°C. Although a large part of China has a cold winter, the temperature rises quickly in spring and is quite high in summer. Furthermore, the rain comes during the warm season, and so rice can be grown across a large part of China during the summer. Even in Xinjiang province and North-East China the temperature is suitable for rice planting in summer. In places south of the Qin Ling Ridge and Huai He River line the conditions are suitable for rice growth for 7–12 months. Here, two crops of rice a year can be grown in quite a wide area, with an additional dryland crop. In South China it is possible to grow three crops of rice a year.

Although much of China's territory is suitable for rice growing, 94 per cent of the planted area is found in southern China, where rice is the dominant crop. This region, which is warm and has a high rainfall, is most suitable for rice growing; here rice can yield more than other crops, and is also more palatable to the local people. Nearly all the rice in southern China is grown by the transplanting method. Rice seedlings are grown on a small portion of land while most of the land is still under wheat, rape, or other crops. As soon as the latter have ripened and been harvested, the rice seedlings in the seed beds are transplanted into the paddy. This practice enables an additional crop to be harvested over most of the region, and it also concentrates the production of rice seedlings into small areas, which is helpful in producing high quality seedlings. The main drawback of the method is that it is labour intensive. However, the population in southern China is dense, and only a small area of cultivated land is available per person — about 0.05 ha in most places, so it is natural that the people there should choose to harvest an additional crop per year at the expense of undertaking a labour intensive activity. This is why the rice paddy in southern China can be used in winter to grow dryland crops like wheat, rape and barley.

In addition a considerable portion of the total paddy area is used in winter to grow winter manure — the Chinese milk vetch (*Astragalus sinicus*) — which is incorporated into the soil at full blossom in the spring to provide a manure for the rice paddy. Irrigation is necessary in order to ensure a good harvest inspite of the high rainfall, since the seasonal distribution of the rain cannot be relied upon. There are, however, some rain-fed rice paddies in the mountainous areas of the Yunnan-Guizhou Plateau in South-West China. A winter crop is not grown on these paddies, instead water is accumulated and stored in the paddy fields for the growth of the rice crop in the following year. The people of southern China have traditionally grown mainly long grain varieties of rice and in recent years there has been a rapid increase in the use of ternary hybrid rice.

In comparison with southern China, rice paddy is sparse in the north, partly due to the poorer climate, but mainly through the lack of water. Here, nearly all the paddies are planted with round grain varieties, and a system of one crop a year predominates, although there are limited areas where two crops a year are grown. Most peasants still use the transplanting method for growing rice, but there are some areas where direct seeding has been adopted. In northern China the transplanting method can be used to extend the growing season because it is possible to take special measures in small-scale seed beds which it would not be possible to take in an open field. The important point about rice growing in northern China is that the crop has to be irrigated, and the irrigation requirement is far greater than that in southern China; it is also two to four times higher than that of wheat, maize, or cotton in North China. This means that it is not economical to plant rice in North China from the point of view of utilizing the available irrigation water, even though the increase in crop yield in response to irrigation is highest for rice. As water resources are in short supply in North China the area planted to rice there has been decreasing in recent years. In some places dryland farming methods have been adopted for rice in order to save irrigation water.

There is a folk song that says that rice tastes pretty good, yet the labour of cultivating rice is quite hard. This is indeed true. Some 60 per cent of the people of China like to eat rice, but growing rice in paddy is exceedingly hard work. Many practices are involved — ploughing, harrowing, levelling, transplanting, weeding, fertilizing, controlling pests, harvesting, threshing, drying, winnowing, and finally taking the grain into the barn. Most of the practices in the field have to be carried out in flooded conditions, and up until now they have been carried out in China mainly by manpower. This means that the peasants have to stand in muddy fields, bending over all day, manually doing this work, whatever the weather. Anybody who has done this work will surely agree with the words of the folk song.

Rice straw, the by-product of the rice harvest, is also a valuable commodity. It can be woven into straw bags, and be used to feed oxen and water buffalo in winter, to thatch houses, and as a raw material for paper. Woven straw shoes were very common in southern China, but are rare now. Rice chaff, the by-product of husking, has long been an important feedstuff for poultry and pigs raised in southern China.

In 1973 the Chinese rice breeder, Yuan Longpin, was the first person in the world to develop ternary hybrid rice, and between 1976 and 1985 a cumulative total of 50.1×10^6 ha was planted with it. This rice is of good quality and it can increase the yield by more than 750 kg/ha. It has contributed considerably to the increase in the rice output of the country in recent years.

Wheat

Wheat is the second largest crop in China, and wheat flour is the staple food preferred by people in northern China. The average area sown to wheat per year in China during 1981–85 was 28.82×10^6 ha, which made China one of the three countries with the largest areas sown to wheat — the others being the USSR and USA.

Wheat is adaptable and is grown nearly everywhere in China, but the main growing area is in North China, where half of the total crop area is located. All of the wheat planted in North China is winter wheat. Because of the short frost-free period in the northern part of North China, it has long been the practice there to inlay crop an autumn-harvest crop into the wheat. In the southern part of North China it is usually possible to grow a second crop after the wheat has been harvested. However, where no irrigation is available it is often necessary to follow the system of three crops in two years since even in the southern part of North China there is insufficient soil moisture available to support two crops per year. In these circumstances, the most usual cropping pattern is to plant a summer crop such as soya bean, millet, or maize after the wheat harvest, and then to plant maize in the spring of the following year. In the western part of North China, especially on the Loess Plateau, there are large areas of dry land, and the most general cropping system there is to harvest four crops every three years. That is, summer fallow is followed by wheat, this is repeated in the next year, then after the wheat harvest in the third year summer millet is multiple cropped with peas or lentils to restore the soil fertility, and the cycle begins again.

Some 30 per cent of the national wheat crop is sown in southern China. This is also winter wheat, and it is all grown in multiple cropping systems. (In China, wheat grown in winter is classified as winter wheat, whether or not the variety requires a cold period to induce flowering, wheat breeders in some other countries would classify this southern-grown wheat as an inter-mediate type). Wheat is one of the crops used in the two crops per year and three crops per year systems for rice paddy, and it occupies the same place in these systems as rape, barley, broad beans, peas, or Chinese milk vetch. When wheat is sown in dryland it is the winter crop of the cropping system of two crops per year with maize or cotton. In southern China wheat is grown mainly in the Yangtze River Valley, with comparatively less wheat being grown in South China. Because the people in southern China prefer to eat rice rather than wheat, a high proportion of the wheat produced is sold. Southern China is, therefore, the main area supplying commodity wheat, although the sown area is only 30 per cent of the total.

Spring wheat is the main type of wheat grown in the areas north and west of North China, and nearly all the cropping follows the system of one crop per year. In the eastern parts of these areas (North-East China and eastern Inner Mongolia) there is some dryland wheat growing due to the moderate rainfall, while in the western parts (Xinjiang and western Inner Mongolia) nearly all the wheat crops are irrigated (the exceptions are in small areas in the Ili He Valley) and therefore the yields are usually higher. The total spring wheat region accounts for one-fifth of the total area sown to wheat in the whole country.

The sowing time for winter wheat in China is from the middle of September to the beginning of October in the northern part of North China, from late September to the middle of October in the North China Plain, from the middle of October to the middle of November in the middle and lower reaches of the Yangtze River, and from the middle of November to the end of that month in the Pearl River Valley (Zhu Jiang River Valley). The sowing time for spring wheat is from the middle of March to the middle of April. Winter wheat is harvested in late April in South China, from the beginning of May to the middle of May in the Yangtze River Valley, from the end of May to the beginning of June in the Huai He River Valley, at the beginning of June in the lower reaches of the Yellow River, and in the middle of June in the northern part of North China. The harvest time for spring wheat is from late July to early August.

The cultivation of wheat is much less arduous than that of paddy rice. It is not necessary to stay in a muddy field for such a long time, and in most cases it is not necessary to bend over while carrying out the cultivation practices. There are also fewer stages in the whole procedure, and most of them do not have to be carried out in the hot summer. In some areas ploughing, sowing, and harvesting have now been mechanized.

The key steps for a good harvest in northern China are the preparation of a good seed-bed and the application of sufficient base manure. If the wheat is irrigated, the irrigations before sowing, before the frosts, and at the time of shooting are the most important. If it is a dryland crop, a complete programme of moisture-conserving tillage practices is necessary early in the spring. For wheat sown in southern China, in addition to applying sufficient base manure, it is also important to dig small, shallow, drainage ditches, as the land may have to be drained during the winter.

Maize

Maize was introduced into China in the sixteenth century. The sown area was rapidly expanded, and maize is now one of the three main crops in the country. In 1986 the area sown to maize in China was 19.1×10^6 ha, and the production was 70.9×10^6 tonnes — an amount second only to that produced in the USA. It is a popular crop, grown over almost all of China, but the main production region is a belt from the North-East, through North China, to Yunnan in the South-West. In areas to the north and west of this belt the climate is not suitable for large-scale planting of maize. In areas to the east and south, rice is the predominant crop, and there is little space left for maize to be grown. Within the maize belt, only a small proportion of the land is planted to rice because of the hilly topography.

Maize production in China can be divided into six different production regions according to the different conditions. The North China Summer Maize Region contains 40 per cent of the total planted area of maize. This is usually summer maize, and it forms the accompanying crop for winter wheat in a two crops per year inlaid cropping or successive cropping system. Part of this region, though, is still planted to spring maize in a three crops every two years system, with winter wheat and summer maize (or other summer crop) as the other crops.

In the Spring Maize Region, north of 40° N, 27 per cent of the national maize crop is planted. Here only one crop per year, of maize, is normally planted, though there is also some intercropping of maize with soya bean. In the South-West Mountainous Maize Region, maize is the dominant summer crop and wheat the dominant winter crop. However, in some places of high altitude spring maize is planted as a single, annual, crop. About 20 per cent of the total area of maize is planted in this region.

The remaining three regions are the Southern Hilly Maize Region, the North-West Irrigated Maize Region, and the Qinghai-Tibetan Plateau Region. They are not as important as the former three regions as together they account for only 13 per cent of the total area planted to maize in the country.

There is not a great deal that should be said specifically about the cultivation of the maize crop, except to emphasize the importance of conserving soil moisture while preparing the seed bed for spring maize in northern China. Since spring drought is a noticeable hazard to cropping in North China, it is necessary to proceed with harrowing straight after ploughing, and to proceed with the 'mu' straight after harrowing, in order to shorten the time during which the soil clods are exposed, and thereby to reduce the evaporative loss of soil moisture. The 'mu' is a tillage implement, and the name is also used for the tillage practice in which the ground is patted down slightly so that the clods are further disintegrated, and the soil surface made more compact without impacting the subsoil. In this way moisture loss from the subsoil is reduced. This practice is equally important for the sowing of other spring crops in North China.

Other 'grain' crops

The other 'grain' crops include barley, sorghum, millet, broom corn millet, soya bean, potato and sweet potato. The sown area of the other 'grain' crops accounted for 27.0 per cent of the total sown area for all 'grain' crops in 1986, and the production of these crops was 14.9 per cent of the total. Clearly these other crops are not as important as rice, wheat, and maize.

Barley, sorghum, millet, and broom corn millet have been cultivated in China for several thousand years. In general, the areas planted to them have been declining because their yields are lower, and/or their tastes are not as good, compared with the three main crops. They do, however, have some specific cultural advantages. Sorghum is tolerant of drought, waterlogging, and salinity, and it can still produce a good yield even after 7 days of flooding to a depth of 30–50 cm of water. The grain of sorghum is an important raw material for making an alcoholic spirit drink called gaoliang jiu.

Millet is characteristically drought tolerant, the seed is particularly suitable for storage, it tastes good, and domestic animals like to eat millet-straw hay. The performance of broom corn millet is similar to that of millet, except that it is even more tolerant of drought, and its growth period is shorter: it takes only 80–90 days to ripen. However, its nutritional value and yield are both lower than for millet.

Barley has a short growing period, is tolerant of low soil fertility, and it is the raw material for malt. The naked barley (*Hordeum vulgare* var. *nudum*), or highland barley, is the staple food of the inhabitants of the Qinghai-Tibetan Plateau, for it can adapt to the conditions of over 4000 m altitude and grow well.

Because of these features there are still some areas sown specifically with these crops. Sorghum, for instance, is sown mainly in North and North-East China, and particularly in North-East China. In 1986, 62 per cent of the total sown area was to be found in Heilongjiang, Jilin, Liaoning, Hebei, and Shanxi provinces, and 70 per cent of the country's total tonnage of sorghum came from these provinces.

Millet and broom corn millet are cultivated mainly in northern China, in areas where irrigation is not possible.

Barley growing is widely scattered, although two-thirds of the crop area is concentrated in the middle and lower reaches of the Yangtze River. The sown area of barley is increasing, due to the development of the beer industry, and the production of barley ranks fourth amongst the other 'grain' crops.

Soya bean originated in China. It was, for a long time, one of the most important export commodities of the country, even in the 1930s China's soya bean production accounted for more than 90 per cent of world production. However, thereafter soya bean production developed rapidly in other countries, while both the sown area and the production of soya bean declined in China. China is now the third largest producer of soya bean: in 1986 8.3×10^6 ha of soya bean were sown, and the total yield was 11.6×10^6 tonnes. Heilongjiang province is the main production area for soya bean in China; its sown area amounted to one-quarter of the national total in 1986, and it producted one-third of the total soya bean crop. Soya bean is grown in Heilongjiang province as a spring-sown, single, annual crop, usually in rotation with spring wheat, sorghum, maize or millet. In other places it is generally sown as one of the two crops in an intercropping system with maize, or sometimes as summer soya bean in a cropping system of two crops a year with wheat. Soya bean has a high content of protein and oil, and it is an important source of these two nutrients for the Chinese people.

According to government stipulation, sweet potato and potato are both designated as 'grain'. In calculating production statistics, 5 kg of these tubers is treated as equivalent to 1 kg of the other 'grains'. The sweet potato is important in China, and accounts for four-fifths of the total area planted to tubers, with potato next. The sweet potato was introduced into China at the end of the sixteenth century and it became a popular crop. Over the last thirty years the planted area has increased gradually. It is a crop worth noticing because of its high yield, and the role it has played in supplying food for the Chinese people, especially during famines, should not be forgotten. The potato is planted mainly in the northern part of North China, and in North-East and North-West China. There are also some other small areas of potato cultivation scattered in other mountainous regions. In most areas of China the potato is used mainly as a vegetable.

Cotton

Cotton is the most important industrial, or cash, crop in China, and its fibre is the main raw material for the textile industry. The area planted to cotton, and the production of cotton, have both increased over the past 30 years. The total area planted to cotton in 1986 was 4.31×10^6 ha, and the total output of ginned cotton 3.54×10^6 tonnes. This was eight times the output of 1949.

Although it is said that cotton was planted about 1500 years ago in Xinjiang province, the cotton now planted was only introduced into China some 600 years ago. In historic times the Chinese peasants used to plant Asiatic cotton (*Gossypium arboreum*) and African cotton (*G. herbaceum*), whose fibres were short and coarse. But at the end of the nineteenth century American upland cotton (*G. hirsutum*) was introduced into China and by the middle of the twentieth century it had completely replaced the former two species. As a result, the staple length of the cotton fibre produced was increased. The long stapled cotton (*G. barbadense*) was introduced into Xinjiang province in the 1950s, and China now produces a small amount of this type of cotton in addition to the main crop.

About half of the total planted area of cotton is in North China. There the natural conditions satisfy the growth requirements of the medium and early maturing upland cotton remarkably well, and the only drawback is a shortage of water during the early stages of growth. In September and October there are many fine days with abundant sunshine, and this is favourable to the late stage of cotton growth and to boll opening. This was once the main production area of cotton. In the early and mid-1950s two-thirds of the total planted area of cotton was planted here, and the yield was one-third higher than in southern China. For a number of reasons, however, the planted area of cotton in this region has decreased considerably since the 1960s, and the yield has increased only marginally, so that in recent years the output of cotton from North China has only amounted to about one-third of the national total. The yield per unit area is now also much lower than in southern China.

Most of the cotton in North China is grown in a system of a single crop per year, continuously planted for quite a long period. However, there are some cotton fields in the southernmost part of the region where inlaid cropping is practised with wheat or oil-bearing crops to obtain two crops per year.

About 40 per cent of the planted area of cotton in China is in the Yangtze River Valley, and the greater part of this is to be found in Jiangsu and Hubei provinces. The average yield of ginned cotton in this region is more than 600 kg/ha, accounting for 60 per cent of the national output. The cotton here is always grown in an inlaid cropping system with winter wheat, and the yield of wheat is generally quite high, so the contradiction between grain and cotton is not obvious in this region in the way that it is in North China. In many places, in the Yangtze River Valley, green manure crops such as burr medic (burr clover, *Medicago polymorpha*), vetch, or broad bean are also inlaid cropped with winter wheat in order to improve soil fertility.

There are also some other areas in northern and southern China where cotton is grown. These are, however, small, and of only minor importance, except for the long-stapled cotton area in Xinjiang province, mentioned previously.

In general, cotton is grown using the direct seeding method. However, in the 1950s the Chinese invented a new cultural method in which a cotton seedling is grown in a nutrient cube and then the cube with the seedling is transplanted into the field. After 30 years' trial with this method a complete cultivation system was evolved and extended to the peasants. In 1980 one-quarter of the cotton sown in the country was planted using this method, and of this 80 per cent was in the Yangtze River Valley where two crops per year are grown. It is obvious that this method of cotton cultivation requires much more labour, but it can bring the growing period forward by 15–20 days, and raise the yield by 10–30 per cent. It also makes it possible to transplant the cotton seedlings after the harvest of the winter wheat crop instead of inlay cropping the cotton into the wheat, which is the only other way of obtaining two crops per year. As there is a surplus labour force in China and land is valuable, the peasants tend to choose this new method of cultivation where it is applicable, despite the additional labour required.

The nutrient-cube system of cotton cultivation is also particularly valuable in areas of high salinity in North China. The seedlings of cotton can be kept alive in highly saline soil if this method is used, and then when the rain comes it leaches the salt from the surface soil into the lower horizons and the cotton is able to grow normally. The result is that these originally worn-out soils can now grow cotton, and the salt content of the soil is, at the same time, reduced.

Other industrial crops

The other industrial crops include the oil-bearing crops, fibre crops, sugar crops, and tobacco. Sometimes silkworm cocoons, tea, fruit, and vegetables are also included in this category.

The main oil-bearing crops in China are peanut and rape (the soya bean is not counted as an oil crop in China), and they accounted for 80 per cent of the production of all oil-bearing crops in 1986. After peanut and rape come sunflower, sesame, and linseed. Peanut is mainly produced in North China, and Shandong province ranks first amongst the provinces and regions where the crop is grown. The total planted area and output of peanut in Shandong province amounted to 26 per cent and 35 per cent, respectively, of the total national figures in 1986. In North China peanut is usually treated as a summer crop, succeeding winter wheat in the cropping system of three crops every two years, though there are still some areas there where a cropping system of two crops per year is practised.

Some 80 per cent of China's rape crop is grown in the Yangtze River Valley, where it is sown as a winter crop, rice being planted in summer. Rapeseed oil is the most popular cooking oil for people living in this part of southern China. The small area of rape grown in North China is sown as a spring crop, generally in a one crop per year system, though some rape is included in the three crops every two years system.

Sunflower is a crop which has only been grown widely in the last 20 years. It is distributed mainly in North-East and North-West China, and in the northern part of North China. Sesame is a popular food with the Chinese and is planted almost everywhere, though the sown areas are all very small. Linseed, conversely, is only grown on the Loess Plateau and in the mountainous areas of North and North-West China. The total sown area is quite small, yet it produces the main cooking oil for the population of these areas.

The fibre crops consist of jute, kenaf or ambari hemp (*Hibiscus cannabinus*), ramie, hemp, flax, and some others. Jute and kenaf are the main fibre crops: in 1986 they accounted for 45 per cent of the area sown to fibre crops and 74 per cent of the total fibre production. Jute is mainly produced in areas south of the Yangtze River, while kenaf is planted mainly in North China. Ramie has the best fibre quality, and cloth woven from ramie fibre has the advantages of stiffness, smartness, ease of drying, and high air permeability; the wearer of ramie cloth feels agreeably cool in summer. Such cloth has for long been a traditional export commodity. Flax fibre is also a good raw material for the textile industry, and linen, together with linen drawn-work and embroideries, are also traditional exports. Most of the flax is grown in Heilongjiang and Jilin provinces in North-East China.

The sugar crops in China are sugar cane and sugar beet. The former is planted in South China and the latter in North-East China and the northern part of North China; the planted areas are quite concentrated. The production of raw sugar in China in 1985 was 4.45×10^6 tonnes, but this was insufficient for domestic consumption and further supplies had to be imported.

Tobacco was introduced into China 300 years ago. In the past the tobacco produced was either sun or air-cured, and flue-curing was only introduced into China at the beginning of the present century. However, the practice spread rapidly and in 1985 tobacco from 82 per cent of the total area planted, and 96 per cent of the tobacco marketed, was flue-cured. Although plots of tobacco are scattered nearly all over the country, the concentrated, and famous, planting areas are the four provinces of Henan, Shandong, Yunnan, and Guizhou.

Tea originated in China. From the fourteenth century to the beginning of the present century it was China's chief export commodity, and Chinese tea occupied an important place in world tea production. Then the tea plantations suffered serious destruction from continuous warfare, but the tea areas have gradually been restored since the 1950s, and in 1976 annual production finally exceeded the highest previous record, achieved in 1886. Production has continued to increase over the last 10 years. Tea is a thermophilic crop which can only grow normally in southern China. The four provinces of Zhejiang, Hunan, Sichuan, and Anhui are the most important tea producing regions, and they account for 60 per cent of the total output. Most of the tea is grown on the hillsides, and the crop occupies little of the cultivated area.

Silkworm cocoons are produced by silkworms raised on mulberry leaves, while the tusser silk is from silkworms raised on oak trees. Silk is, of course, a famous Chinese product, and the history of its production is very long. There was serious destruction of silk production during the Japanese invasion, but production is now restored. The quantity of raw silk exported from China has recently amounted to 90 per cent of the total world trade, while the exports of silk and satin cloth have comprised some 40 per cent of the total world trade in these items. The most important production areas for silkworm cocoons are the Yangtze River Delta, the Zhu Jiang (Pearl) River Delta, and the Chengdu Plain; while the most important area for tusser cocoons is Liaoning province. The tusser cocoon output of Liaoning province accounts for 80 per cent of the national total, and the next most important producing areas are in Henan and Shandong provinces.

The most important fruits of China are apples, pears, and oranges, and in 1986 they accounted for 25 per cent, 17 per cent, and 19 per cent, respectively, of total fruit production. Apples are produced mainly in the eastern part of North China and the southern part of North-East China, and Shandong and Liaoning provinces are the largest producers. Pear trees are distributed quite widely, though the main growing area is still in North China, especially Hebei province. Orange trees are more thermophilic, and they are only grown in southern China as they can not survive the winter of northern China. The main orange producing areas are rather scattered, but there is a comparative concentration of production in Sichuan, Guangdong, Hunan, and Zhejiang provinces. There are many other types of fruit to be found in China, but the amounts produced are quite small. Generally speaking, the amount of fruit that ordinary Chinese people eat is also quite small; only those people who live in the production areas, or urban dwellers, are able to eat a little more.

By comparison, the amount of vegetables eaten is much larger. The Chinese diet consists mainly of grain and vegetables with only a little meat and practically no milk products — except amongst those minority peoples who live on animal husbandry. A large proportion of the Chinese people can endure eating no meat for quite a long period without difficulty, yet they would feel rather unwell if for some days there were no vegetables in their meals. Chinese peasants always plant vegetables in their land; and the extent of vegetable growing around big cities is always very large. In 1986 there were 5.30×10^6 ha sown to vegetables in the whole country, with 58 000 ha of this in the municipality of Beijing, and 71 000 ha in the municipality of Shanghai.

Green manure and green fodder crops

There is a long tradition in southern China of growing a green manure crop, such as milk vetch, vetch, or broad bean, in the rice paddy in the winter. In general, the crop is sown in October and is ploughed into the ground at full bloom stage in April of the following year. But the total area of green manure has decreased considerably in the last 30 years as a result of the one-sided emphasis on 'grain' production. For cotton, burr medic (burr clover) is the main form of green manure in southern China, while in North China mung bean is occasionally sown after the wheat harvest as a green manure crop. There are also some areas where lucerne (alfalfa) is grown, but this is mainly grown as feed for domestic animals, particularly horses and mules, rather than for green manure. Some sesbania is planted in North China, particularly where there is a salinity problem, as it has a high salt tolerance.

The national statistics report 4.44×10^6 ha planted to green manure crops in 1986, and a further 1.74×10^6 ha planted to green fodder crops. The largest total areas of green manure crops were planted in Hunan and Jiangxi provinces, and the next largest in Hubei, Zhejiang, and Jiangsu, in that order. All these provinces are in the middle and lower reaches of the Yangtze River Valley, and together they accounted for 77 per cent of the total national area sown to green manure. The largest total areas of green fodder crops were planted in Sichuan and Gansu provinces.

Forestry

It is estimated that there are more than 5000 species of trees in China, and many of them are precious, such as the metasequoia (*Metasequoia glyptostroboides*) and the ginkgo (*Ginkgo biloba*). However, there are not many forests in China. According to the State Statistical Bureau, China's forest coverage was 12 per cent in 1984; this is not large. Half of the total forest area and three-quarters of the timber reserves are concentrated in two small areas in remote North-East and South-West China, which makes the forest in other places even more sparse. There are only a few places in the extensive western half of China where trees grow, and the forest coverage amounts only to about one per cent. There are natural, historical, and social reasons for this. The natural distribution of forests with tall timber trees is restricted in general to areas within the 400 mm isohyet, and almost all of the land within the western part of China lies outside this boundary. Indeed, in much of the area the rainfall is less than 250 mm, and many places receive less than 100 mm, so it is not possible to have forests of tall trees.

There is no doubt that in ancient times the eastern part of China had plenty of dense forests. But the development of cultivation and the growth of population, together with the effects of wars and disasters, combined to bring about the destruction of the original forests, and the formation of wasted, rocky, mountain-landscapes.

The Chinese government has attached great importance to reafforestation in the past 30 years, yet the result has not been satisfactory for various reasons. One such reason is that the Chinese people are in the habit of eating hot meals, and this leads to the consumption of a large amount of firewood, given that there are more than 800 million people living in rural China. It is reported that the total area reafforested within the last 30 years has amounted to 100 million ha, but the woods have actually been kept alive in less than 30 per cent of this area.

During the same period the timber felling capability has increased greatly, and the timber output has increased more than nine times. The increase in timber output is an advantage, provided that it is within the scope of the growth capability of the forests; if it is not, it will be a decided disadvantage. In fact, there has already been a warning that if the present phenomenon of timber being felled faster than it grows is not eliminated, there will soon be no timber available. A further warning has also been given that if the over-felling in some areas is not stopped, serious erosion will appear, and there will be further harmful effects on the local rivers and climate. The Chinese government is aware of this situation and has given great importance to the promotion of reafforestation and grass cultivation. As the time of tree planting approaches in spring or autumn, the Chinese Communist Party and government leaders always come out to plant trees, and the newspapers and the radio broadcasters call on the people to plant trees too. In 1978 the government began to plant wind-break forests in the 'Three Norths', that is North-East, North, and North-West China. The intention is to plant a wide range of forests, and it was reported that the first stage of this work was completed at the end of 1985, more than doubling the forest coverage, to 5.9 per cent. This is a project on an enormous scale, and it will have a major effect after it is completed.

Because of the comparative lack of timber resources, China has long been an importer of timber for construction, and of wood pulp and chip for paper making. Most of the Chinese people plant several trees around their house, and some trees around their village. If there are hills around a village, the inhabitants will plant a small area of woods to provide building material and firewood. This is carried out in the peasants' spare time as a sideline activity; it does not occupy an important position in the rural economy. All of the woods and forests belonged to the State from 1949, and were managed by government forestry units until the beginning of the 1980s. More recently, many worn-out hills and waste lands have been contracted out to peasants for them to plant woods, and a person who plants a wood under the scheme can enjoy the ownership of that wood for at least 50 years. However, it will take time for the trees to grow.

Animal husbandry

The archaeologists believe that animal husbandry has been practised in China for at least the last seven or eight thousand years. Bones of pigs, dogs, and goats were found in the Peiligang ruin, in Henan province, which has been dated to 8000 B P by radio-carbon dating, and in the Hemudu ruin, Zhejiang province, which has been similarly dated to 7000 BP. However, apart from the nomadic minorities in western China, which amount to a very small proportion of the country's total population, the Chinese people are cultivators by nature and their animal husbandry, on the whole, is a minor activity. Animal production has been developed over the last 30 years, though, and the output of animal products now accounts for one-fifth of the Gross Output Value of Agriculture (see Table 3.1); however, this output is still small compared with that of crop cultivation.

According to Chinese practice, the western part of China (corresponding to the Xinjiang-Inner Mongolia Arid Region plus the Qinghai-Tibetan Plateau Region of Chapter 1) is called the pastoral region, and the eastern part (the Eastern Monsoon Region), the agricultural region. There are immense areas in the western part of China which are not suitable for crop cultivation because of their dryness or of the altitude. However, historically, this has long been the place where the Chinese minorities live, whose habits and customs are distinctly different from those of the peasants who live in the eastern part of China where the Han nationality is dominant. The result, then, is the formation of two different landscapes, two different cultures, and different animal husbandries.

Animal husbandry in the western part of China is managed by people of the minority nationalities who raise sheep and cattle, with some horses, and also camels in some places. They keep their animals primarily to provide items for self-consumption, and meat eating and milk drinking are part of their life style. These people live a nomadic life, travelling on horseback with their flocks and herds, and taking their tents with them, as they seek out the best grazing and available water. Traditionally, they do not settle in one place or cultivate crops.

In the eastern part of China, on the other hand, animal husbandry is managed by the peasants of Han nationality, who keep pigs and poultry in pens. The purpose of keeping animals and poultry is to provide some meat and eggs to eat with rice or bread — but not to provide the main food. The peasants in the eastern part of China also keep oxen or horses, but these are kept for draught purposes, not for meat.

Although the western part of China is the pastoral region, crop cultivation has now been developed there to a considerable extent. The Gross Output Value of the animal products (including those consumed by the producer) produced in each of five western provinces and autonomous regions in 1986, together with the percentage which this value represented of the total Gross Output Value of Agriculture is shown in Table 3.4. Even in Tibet and Qinghai, the two provinces or regions with the highest proportions of animal production in the total agricultural production, the value of the animal products only just exceeded that of the crop products in the former, and did not do so in the latter. In Inner Mongolia animal production amounted to less than one-third of total agricultural production, and in Xinjiang and Ningxia the proportion was less than one-fifth.

Animal husbandry is now conducted on quite a large scale in the eastern part of China, and the production figures for the five provinces with the highest levels of animal production in eastern China in 1986 are also given in Table 3.4. It can be seen that the contribution of animal husbandry to total agricultural production in all but one of these provinces was greater than 20 per cent, and greater than that in Xinjiang.

Table 3.4 also shows that the total value of animal produce in the five

TABLE 3.4   The Gross Output Value of Animal Husbandry (GOVAH) in selected provinces and autonomous regions at current prices in 1986

Province or autonomous region	GOVAH (10^8 yuan)	GOVAH as a percentage of GOVA (%)
Western part of China		
Inner Mongolia	23.99	31
Xinjiang	13.36	20
Qinghai	6.11	43
Tibet	5.36	54
Ningxia	2.82	20
Total	51.64	
Eastern part of China		
Sichuan	94.08	28
Jiangsu	69.47	21
Shandong	62.82	17
Guangdong	68.41	22
Hunan	58.40	26
Total	353.18	

Source: Editorial Board of China Agriculture Yearbook (1988). *China Agriculture Yearbook 1987 (English Edition)*. Agricultural Publishing House, Beijing.

western pastoral provinces and autonomous regions taken together, amounted to 5164×10^6 yuan: a lower figure than for any of the five eastern provinces taken separately. Clearly, from the point of view of production value, the scale of animal husbandry in the western pastoral region is far smaller than that in the eastern part of China. The total Gross Output Value of Animal Husbandry for the whole country in 1986 was $87 354 \times 10^6$ yuan, and of this the five western provinces listed in Table 3.4 contributed 5.9 per cent, and the five eastern provinces 40.4 per cent.

Given that animal production is larger in the eastern part of the country than in the west, it is easy to accept that the main animal products are pork, poultry meat, and eggs. As shown in Table 3.5, pork represented a very major portion of total animal production in 1986. Beef and mutton taken together only made up 5.7 per cent of total meat production, and milk production was 60 per cent of egg production by fresh weight. Pig and poultry production therefore dominate Chinese animal husbandry.

The main five producing provinces or autonomous regions for selected animal products in 1986 are shown in Table 3.6. Sichuan province had the largest total output of animal products and was also the largest producer of meat in total, and of pork. The next four provinces with the largest outputs by these measures (Hunan, Jiangsu, Guangdong, and Shandong) were the same, just the order varied a little according to the parameter. This, of course, reflects the dominant position of pig breeding in animal husbandry.

Provinces or autonomous regions in the western pastoral region ranked highly as producers of beef, mutton, and milk, reflecting the dominance of sheep, goat, and cattle husbandry in this region, and the lesser interest in these forms of animal husbandry in eastern China. Furthermore, within the eastern part of China, the provinces and autonomous region with the greatest outputs of pork, poultry meat, and eggs were mostly in southern China, except for Shandong and Hebei; but mutton and beef were produced mainly in the northern provinces.

TABLE 3.5   The main animal products of China in 1986

Product	Quantity (10^3 tonnes)	Product	Quantity (10^3 tonnes)
Pork	17 960	Cow's milk	2899
Poultry meat	1879	Sheep and goat milk	430
Sheep and goat mutton	622	Sheep wool	185
Beef	589	Goat wool	12
Rabbit meat	74	Cashmere	3
Eggs	5550	Rabbit wool	3
Honey	172		

Source: Editorial Board of China Agriculture Yearbook (1988). *China Agriculture Yearbook 1987 (English Edition)*. Agricultural Publishing House, Beijing.

TABLE 3.6   The provinces or autonomous regions which were the top five producers of selected animal products in 1986

| Product | Rank order of producing provinces or autonomous regions | | | | |
	First	Second	Third	Fourth	Fifth
Total meat	Sichuan	Hunan	Jiangsu	Guangdong	Shandong
Pork	Sichuan	Hunan	Jiangsu	Shandong	Guangdong
Poultry meat	Guangdong	Jiangsu	Sichuan	Anhui	Guangxi
Sheep and goat mutton	Xinjiang	Inner Mongolia	Shandong	Qinghai	Hebei
Beef	Henan	Shandong	Inner Mongolia	Xinjiang	Sichuan
Rabbit meat	Sichuan	Shandong	Hebei	Jiangsu	Shanxi & Zhejiang
Eggs	Shandong	Jiangsu	Hubei	Henan	Hebei
Milk	Heilongjiang	Inner Mongolia	Sichuan	Xinjiang	Shaanxi & Tibet

Source: Editorial Board of China Agriculture Yearbook (1988). *China Agriculture Yearbook 1987 (English Edition)*. Agricultural Publishing House, Beijing.

Table 3.7 gives the numbers of livestock for the years 1952–86. The figures are for the numbers of animals alive at the end of each year, which is the usual convention for showing the magnitude of animal husbandry in China. This is not a precise method since it does not reflect the numbers of animals slaughtered or sold within the year, but it does give an approximate magnitude. During the period 1952–86 pigs showed the greatest increase, their numbers rising by a factor of 3.8. There was a three-fold increase in the number of mules, and sheep and goats together increased 2.7 times. The remaining domestic animals showed a less than two-fold increase in numbers, apart from the donkeys which showed a slight decrease. In 1952, pork, beef, and mutton production amounted to 3.385×10^6 tonnes, this increased to 19.171×10^6 tonnes in 1986, an increase of 5.7 times. Thus the large increase in animal husbandry in the last 30 years or so has been mainly due to the rapid development of pig husbandry, together with the expansion of sheep and goat husbandry. However, figures for the poultry industry are not included here, and it should be noted that this also developed over the same period to a considerable extent. In recent years factory farming of pigs and poultry has been started in some of the big cities, and also the breeding of lean pigs for pork.

Chinese animal husbandry faces various problems. Firstly, there is a shortage of feed and/or a problem of poor feed quality. The grass grown in the western pastoral region was always poor, and has been getting worse as the numbers of animals have increased in recent years, leading to over-stocking in some places. This has decreased forage quality, accelerated desertification, and caused a rise in the incidence of damage by wild animals, including rodents. It is a serious problem which requires attention. Finding

TABLE 3.7   The total numbers of various livestock at the year end, 1952–86

Year	Pigs (10^6)	Sheep & goats (10^6)	Cattle & buffalo (10^6)	Horses (10^6)	Donkeys (10^6)	Mules (10^6)	Camels (10^6)
1952	89.77	61.78	56.60	6.13	11.81	1.64	0.29
1957	145.90	98.58	63.61	7.30	10.86	1.68	0.37
1965	166.93	139.03	66.95	7.92	7.44	1.45	0.45
1970	206.10	147.04	73.58	9.65	8.40	2.25	0.49
1975	281.17	163.37	73.55	11.30	8.13	3.35	0.54
1980	305.43	187.31	71.68	11.04	7.75	4.17	0.61
1985	331.40	155.88	86.82	11.08	10.42	4.97	0.53
1986	337.19	166.23	91.67	10.99	10.69	5.11	0.50

Source: Editorial Board of China Agriculture Yearbook (1986). *China Agriculture Yearbook 1985 (English Edition)*; Editorial Board of China Agriculture Yearbook (1988). *China Agriculture Yearbook 1987 (English Edition)*. Agricultural Publishing House, Beijing.

sufficient feed of adequate quality for pig and poultry production in the eastern part of China is also a major problem. Not only is there insufficient feed but its quality is poor, there is a lack of variety, and, most serious of all, the supply of protein for feed processing is far from sufficient. The further advancement of animal husbandry in China will depend upon overcoming these deficiencies.

The second problem is that the quality of the animal and poultry breeding stock is poor, leading to low productivity. It is not uncommon for a hog to take 18 months to reach slaughter weight at 90–100 kg, and most fowls only lay 100–150 eggs per year. This level of productivity makes the cost of production high, and thereby makes it difficult for people running an animal or poultry enterprise to make a profit. Consequently, enthusiasm for running such enterprises is often lacking.

Thirdly, the methods of animal rearing need to be improved. Although the Chinese have a very long history of animal husbandry, the keeping of animals has usually been a matter for individual families, and has been treated as a minor activity. This is different to many modern ideas of animal husbandry. For instance, Chinese peasants do not pay much attention to feed efficiency, economic return, or the methods of breeding. They often incorporate large amounts of green feed with a small amount of grain to feed the pigs. The assumption is that this practice will lower grain consumption, and, therefore, cost; in reality it prolongs the growing period of the pigs to such an extent that grain is wasted and the production cost is increased. In addition, hogs fed mainly on green feed during the early stages of growth and then fattened on grain, have more fat and less lean meat, have rough muscle fibres, and do not command a good price.

The value of the animal products produced for the Chinese population in 1986 was about 83 yuan per person, and the quantity of meat produced (including poultry and rabbit meat) was about 20 kg per person; both figures were rather low. It is certain that animal husbandry in China will develop further in the future, and it is equally certain that the problems mentioned above will have to be solved sooner or later.

The fishery

China is bordered on the eastern side by the Pacific Ocean and along this coast there are various local waters which cross the temperate, subtropical, and tropical zones, covering almost 40 degrees of latitude. From north to south, these are the Bo Hai Sea, the Yellow Sea (Huang Hai), the East China Sea (Dong Hai), the South China Sea (Nan Hai), and the Pacific Ocean east of Taiwan. The coastline of the mainland starts at the Yalu Jiang River in Liaoning province in the north, and goes down to the mouth of the Beilun He River in Guangxi Autonomous Region in the south — a length of 18 000 km.

If the coastline of the 5000 islands is included, the total coastline of China amounts to 32 000 km. The area of the continental shelf fishing ground, within the 200 m isobar, is 1.5 million km^2, and there are 1.3 million ha of shallow water and shoals which can be developed into fish farms for marine aquaculture. Because most of the rivers of China flow eastwards into the seas of the Pacific Ocean, there are plenty of habitats for marine animals and plants along the sea coast of China.

There are about 20 million ha of inland water including lakes, ponds, reservoirs, rivers, and streams, of which 5 million ha may be utilized for aquaculture.

The Chinese people have a long tradition of catching and raising fish: methods of fish raising and principles of fish catching were recorded 2000 years ago. However, for various reasons, the fishery output in 1949 was only 450 000 tonnes. After the founding of the People's Republic of China, the government implemented a series of policies to promote and support the development of the fishery, and thenceforth aquatic production was gradually increased, reaching 3.12×10^6 tonnes in 1957, an increase of 6.9 times over the 1949 level. Unfortunately, in succeeding years there was little further increase in production, due to political as well as administrative factors. In 1979 the situation changed again with the advent of economic reform; from then on reasonable utilization of resources and the development of aquaculture were emphasized, and efforts were made to improve the quality of the produce. At the same time the production responsibility system was introduced (see Chapter 6), the results of research were communicated to the fishermen and fish farmers, and steps were taken to protect the fish resource, especially fish reproduction (Table 3.8).

The marine fishery is an important part of the Chinese fishing industry, and the annual marine catch amounts to about half of the output of the total Chinese fishery. There are more than one million fishermen and administrative staff involved in the marine fishery, and over 100 000 motorized fishing boats of about 40 hp on average. Apart from about 1000 fishing boats owned by 40 state-run enterprises, most fishing boats belong to collective units, while some are managed by individual fishermen. There are 700 fishing harbours of various sizes along the coast, 30 of which have been equipped with comparatively advanced facilities and which serve as marine fishing bases. In general, they all have a meteorological observatory, a sheltered anchorage, fish freezing facilities, provision for the supply of materials, storage facilities, and transport facilities, together with an accompanying infrastructure of such facilities as hospitals and schools.

The fishing methods used in the marine fishery include trawling, purse-seining, gill-netting, set-netting, and lining, with trawling as the main method. In the past there was an excessive development, of motorized trawling boats, which were mainly operated in offshore waters and without strict management. As a result some species of marine fish were exhausted

TABLE 3.8 The output of aquatic products in China, 1950–87

Year	Total output	Marine products		Freshwater products	
	(10^4 tonnes)	Caught (10^4 tonnes)	Cultured (10^4 tonnes)	Caught (10^4 tonnes)	Cultured (10^4 tonnes)
1950	91	54	1	30	6
1955	252	155	11	54	32
1960	304	175	12	67	50
1965	298	191	10	46	51
1970	318	210	18	32	58
1975	441	307	28	31	75
1980	450	281	45	34	90
1985	705	349	71	47	238
1986	824	389	86	54	294
1987	955	438	110	60	347

Source: Editorial Board of China Agriculture Yearbook (1986). *China Agriculture Yearbook 1985 (English Edition).* Agricultural Publishing House, Beijing; State Statistical Bureau (1988). *China Statistical Yearbook 1988.* International Centre for the Advancement of Science and Technology Ltd, Hong Kong, and China Statistical Information and Consultancy Service, Beijing. (In English)

and this seriously influenced the development of China's marine fishery. Since 1978 the Chinese government has enhanced the management of the fishery and imposed strict discipline on the areas and seasons where and when fishing is prohibited, and also restricted the use of harmful fishing gear. As a result the marine fishery resource has been protected and has shown some recovery (Table 3.9).

China's pelagic fishing fleet began to form in 1985, when Chinese fishermen first appeared in the waters west of Africa, in the Bering Sea, and in the Gulf of Alaska. Since then various forms of fishing cooperation with foreign enterprises have been established. China is now expanding her pelagic fishing fleet in order to develop further fishery cooperation with other coastal countries of the world (Table 3.10).

China's marine culture has made rapid technological progress in recent years: the artificial hatching of many species has been achieved, cultural techniques have been improved, and the unit area yield has increased year by year. The species cultured have been increased from the traditional shellfish in the shoals along the sea-shore, such as oysters, razor clams, and ark shell clams, together with laver, to include such species as kelp, blue mussels, prawns, bay scallops, sea-cucumbers, abalone, and pearl mussels. Altogether 30 species are now cultured. Before 1980 the production of algae occupied a very large proportion of total production; however, the production of fish, shrimp and other valuable species is gradually increasing.

TABLE 3.9   The output of selected marine products in China, 1957–87

Year	Big yellow croaker (10⁴ tonnes)	Small yellow croaker (10⁴ tonnes)	Hairtail (10⁴ tonnes)	Scad & mackerel (10⁴ tonnes)	Jellyfish (10⁴ tonnes)	Kelp (10⁴ tonnes)
1957	17.8	16.3	20.0	1.2	3.3	0.6*
1960	6.6	13.6	28.0	1.1	5.9	4.9
1965	10.3	4.4	37.8	3.5	2.6	2.7
1970	15.9	3.0	39.2	17.3	5.1	8.8
1975	14.0	5.5	48.4	8.4	1.7	16.0
1980	8.6	3.6	47.3	24.7	0.9	25.3
1985	2.6	3.1	45.9	32.7	6.1	25.4
1986	1.7	2.0	40.6	37.0		
1987	1.7	2.0	39.3	51.1		

* 1958 figure.
Source: Editorial Board of China Agriculture Yearbook (1986). *China Agriculture Yearbook 1985 (English Edition)*. Agricultural Publishing House, Beijing; State Statistical Bureau (1988). *China Statistical Yearbook 1988*. International Centre for the Advancement of Science and Technology Ltd, Hong Kong, and China Statistical Information and Consultancy Service, Beijing. (In English)

About half of the total population engaged in the marine fishery operates in the waters south of the Yangtze River, and 40 per cent of the total marine production comes from this area. Here, too, is found the biggest fishing ground, the Zhoushan fishing ground, in Zhejiang province, and also the largest market for marine products, and the centre of fish processing, both of which are in Shanghai. Shanghai is also the most important base for scientific research and education for the marine fishery.

The marine fish production of the coastal provinces of China is shown in Table 3.10. In 1987 Zhejiang province was the largest producer of marine products, mainly because of its high fishing catch, while the leading producers of cultured products were Liaoning, Shandong, and Fujian provinces.

China's output of freshwater fish ranks first in the world. There are dozens of species of fish and shrimps suitable for raising, and there are 5 million ha of freshwater rivers, lakes, reservoirs, and ponds available in which fish can be raised. Many new techniques have been developed in recent years for use in the freshwater fishery. Artificial propagation, cross breeding, mixed culture of different species, high intensity pisciculture in running water, net-cage pisciculture, fish disease control, the use of compound fish feeds, and the use of oxygen charging apparatus, have all led to substantial yield increases. The places with the longest history of freshwater pisciculture, and with the most advanced technology, are in the

TABLE 3.10   Marine fishery production of the coastal provinces of China, 1987

Province or municipality	Total output (10^3 tonnes)	Caught (10^3 tonnes)	Cultured (10^3 tonnes)
China total	5481.6	4380.8	1100.8
Liaoning	759.2	437.7	321.5
Beijing	0.2		0.2
Hebei	140.6	114.4	26.2
Tianjin	32.9	28.1	4.7
Jiangsu	279.8	250.6	29.2
Shanghai	176.1	174.1	2.1
Shandong	983.1	717.6	265.5
Zhejiang	1028.4	909.8	118.6
Fujian	910.7	684.4	226.4
Guangdong	1001.0	896.0	105.0
Guangxi	151.4	150.1	1.3

Note: The catch of the ocean-going fleet of the China National Fisheries Corporation was 18 069 t in 1987, and this is included in the above total for China.
Source: State Statistical Bureau (1988). *China Statistical Yearbook 1988*. International Centre for the Advancement of Science and Technology Ltd, Hong Kong, and China Statistical Information and Consultancy Service, Beijing. (In English)

counties of Shunde, Zhongshan, and Nanhai in Fushan prefecture, Guangdong province; and in the counties around Lake Tai Hu — that is, the counties of Wuxi and Suzhou in Jiangsu province, and of Huzhou and Jiaxing in Zhejiang province.

China's freshwater aquaculture is still expanding. The main tasks in the newly developed areas are to increase the cultivated area, raise the yield per unit area, improve the quality of the aquatic products, and increase the efficiency of production, whereas in the old aquaculture areas the main needs are to increase the varieties used and extend intensive culture. The practice of using enclosures or net-cages for fish cultivation in reservoirs, lakes, or other waters is becoming popular. Many major fish producing regions are attempting to increase their proportion of high quality fish in order to try to meet the ever growing demand in the market which has resulted from economic reform. For the same reason a rotational system of catching and stocking is also being followed so that a steady supply of aquatic products can be provided. Table 3.11 shows the freshwater aquatic production of the ten provinces which had an output in excess of 100 000 tonnes, in 1987. The first eight of these provinces are in southern China. The combined output of these 10 provinces was 3.4×10^6 tonnes, which represented 84 per cent of the total for the whole country.

TABLE 3.11 Freshwater aquatic production of provinces with an output in excess of 100 000 tonnes in 1987

Province	Total output	Caught	Cultured	Of total output		
				Fish	Shrimps, prawns, and crab	Shellfish
	$(10^3$ tonnes)	$(10^3$ tonnes)	$(10^3$ tonnes)	$(10^3$ tonnes)	$(10^3$ tonnes)	$(10^3$ tonnes)
China total	4071.9	599.5	3472.4	3938.0	80.5	53.3
Guangdong	665.0	31.0	634.0	654.0	3.0	8.0
Jiangsu	642.1	128.0	514.1	609.1	19.0	14.1
Hubei	582.7	69.4	513.3	561.6	13.9	7.2
Hunan	443.8	37.6	406.2	434.1	3.5	6.2
Anhui	240.7	59.3	181.4	223.2	12.9	4.6
Jiangxi	225.9	40.2	185.7	218.8	3.8	3.3
Zhejiang	221.5	25.0	196.5	217.0	1.9	2.6
Sichuan	171.7	23.5	148.2	171.0	0.7	–
Shandong	123.5	34.1	89.4	111.5	9.0	3.0
Heilongjiang	101.5	32.1	69.4	99.7	1.2	0.6

Source: State Statistical Bureau (1988). *China Statistical Yearbook 1988*. International Centre for the Advancement of Science and Technology Ltd, Hong Kong, and China Statistical Information and Consultancy Service, Beijing. (In English)

Although China has a long tradition of processing aquatic products, only simple methods were used before the 1950s. These were salting, drying, pickling with grains and wine, saturating with liquid, and smoking. After 1949 the manual workshop type of production was gradually replaced by industrial production on a comparatively large scale. At present there are 430 processing factories in China, with more than 50 000 workers and management staff. Of these, 240 are state-run factories which have an overall capacity to process about two million tonnes of fresh fish annually. There are 370 cold storage units for aquatic products with a total storage capacity of 250 000 tonnes/batch; a freeze capacity of 8000 tonnes/day, an ice making capacity of 7000 tonnes/day, and an ice storage capacity of 190 000 tonnes. These processing factories and cold storages are spread over the key fishing areas along the coast and inland and make up a network of facilities for handling aquatic products. They produce more than 2000 types of products which can be grouped into five categories — frozen (72 per cent of the total output), canned (8 per cent), hard cured (13 per cent), cooked (4 per cent), and comprehensively utilized products (3 per cent). Some new types of fish food have been developed in recent years, such as those made of ground fish meal, imitation aquatic food, and aquatic foods with a traditional flavour. Apart from supplying a large amount of aquatic food, the aquatic processing industry also supplies substantial quantities of raw materials, by-products, and semi-processed goods to the medical, chemical, food processing, printing and dyeing, textile, military, and light industries.

Because of the large population, the per caput production of aquatic products for the Chinese people is low, it amounts only to about 9 kg/year, and the supply of aquatic products is always insufficient for the requirements of the domestic market. There are, in fact, quite a large number of people who live in the mountainous areas inland, especially on the North-West Plateau, who eat little aquatic food. However, in some places in the coastal provinces the situation has changed since the change of policy at the beginning of the 1980s. In recent years the prices of aquatic products have been adjusted, fish raising is encouraged, direct salting by fishermen is permitted, and a simplified handling and marketing system has been introduced, all of which has resulted in a better supply of aquatic products to the market.

The main exports of aquatic products from China are of prawns and shrimp to Japan, and of live freshwater fish to Hong Kong. In recent years the total value of these exports has reached about US$300 million annually.

References

The following texts are in Chinese:

Division of Economic Geography, Institute of Geography, Chinese Academy of Sciences (1983). *The Production Pattern of Chinese Agriculture*. Agricultural Publishing House, Beijing.

Editorial Group of the Plant Industry Regionalization of China (1984). *The Plant Industry Regionalization of China*. Agricultural Publishing House, Beijing.

Shen Xuenian, Liu Xunhao, and others (1983). *Multiple Cropping*. Agricultural Publishing House, Beijing.

The following texts are in English:

Editorial Board of China Agriculture Yearbook (1986). *China Agriculture Yearbook 1985 (English Edition)*. Agricultural Publishing House, Beijing.

Editorial Board of China Agriculture Yearbook (1987). *China Agriculture Yearbook 1986 (English Edition)*. Agricultural Publishing House, Beijing.

Editorial Board of China Agriculture Yearbook (1988). *China Agriculture Yearbook 1987 (English Edition)*. Agricultural Publishing House, Beijing.

4

The ten agricultural regions of China[1]

YANG SHENGHUA

The Regionalization of agriculture in China

The different agricultural regions of China, and the problems facing them, will be described in this Chapter. The regionalization system for agriculture now used widely in China is the system proposed by the National Committee for Agricultural Regionalization and published in 1981. This divided the whole country into 10 agricultural regions as follows (see Fig. 4.1):

I.   North-East China
II.   Inner Mongolia and Areas Along the Great Wall
III.   North China Plain
IV.   Loess Plateau
V.   Middle and Lower Yangtze River Valley
VI.   South-West China
VII.   South China
VIII.   Gansu and Xinjiang
IX.   Qinghai and Tibet
X.   The Region of Sea Fishery

According to *A Comprehensive Agricultural Regionalization of China*, 1981 (in Chinese), which is the report of the National Committee for Agricultural Regionalization, the 10 agricultural regions mentioned above are not regionalized at only one level. The Committee concluded that the greatest agricultural regional difference is that between the eastern part and the western part of China. Throughout the eastern part there is a similar combination of solar energy, water resources, and soil, together with the people's long historical experience of agriculture, and a dense population. Most of the cultivated land, crops, forestry, fishery, and sideline activities, are concentrated here. In contrast, the western part is more or less arid, with

---

1. The figures quoted in this chapter are from *A Comprehensive Agricultural Regionalization of China*, 1981, and are derived from the statistics of 1978. Statistics on the basis of the agricultural region boundaries are not otherwise readily available. The yield figures are per hectare of cultivated land per year unless otherwise indicated, and the agricultural population is as defined in Appendix 3.

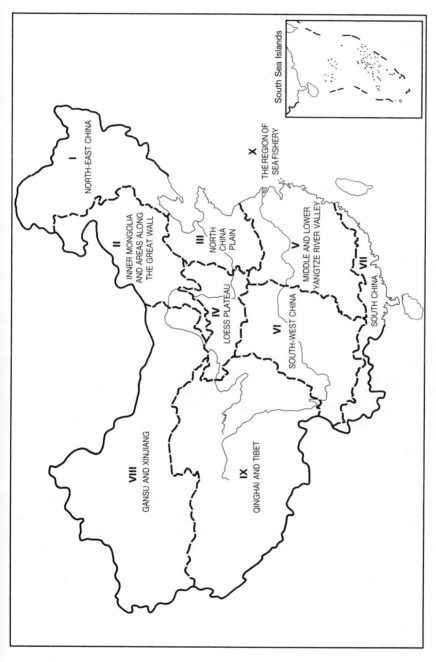

FIG. 4.1   The agricultural regions of China.

a less favourable combination of solar energy, water resources, and soil. It is where the national minorities live, and the history of cropping there is shorter. The population in the western part is quite sparse, cropping is generally in isolated small areas, and most of the land is used for grazing.

A second division can also be made, in which the eastern and western parts are each divided into northern and southern parts. In the east, the dividing line between northern and southern China is the Qing Ling Ridge–Huai He River line. In southern China rice paddy is the dominant land use, and in northern China dry-crop cultivation predominates. In western China, the Kunlun Shan–Qilian Shan Ranges divide the arid northern Gansu and Xinjiang Region (Region VIII) where cropping is nearly impossible without irrigation, and the extensive desert and mountainous areas are used for grazing, from the southern Qinghai and Tibet Region (Region IX), of the famous Qinghai–Tibetan Plateau. Here the altitude is high and the climate cold, grazing is the main agricultural activity, and there is little cropping.

Northern China is further divided to give four subdivisions. North-East China (Region I) is a region with a bitterly cold winter, although summer is warm. One crop is grown per year, mainly 'grain', and a surplus over local requirements is produced. The history of cropping here is comparatively short, and some wasteland suitable for arable use is still available for reclamation. There are also good forest reserves in the region. Inner Mongolia and Areas Along the Great Wall (Region II) constitutes a transitional zone between the south-east monsoon area and the north-west arid plateau. It is the least populated region in the eastern part of China, and the farming activity is mainly grazing with some cropping (one crop per year) mixed with the grazing in the south. The North China Plain (Region III) is densely populated and has a large area of cultivated land. There is a long history of cultivation, and the cropping index is quite high. The dominant cropping system is two crops per year or three crops in every two years. It is the main production area for 'grain' (except rice), cotton, and oil seeds. The Loess Plateau (Region IV) is essentially an area of plateau and hilly land covered with loess layers of varying thickness. Most of the region has a semi-arid climate. Dry farming is important, and 'grain' crops with a high drought tolerance are grown in a system of one crop a year or three crops in every two years. Crop yields are rather low and are not reliable, and soil erosion is a serious problem, except in the south-eastern part. In general, the people living here have a hard life.

Southern China is divided to give three subdivisions. The Middle and Lower Yangtze River Valley (Region V) is situated in the subtropics, with plains, rolling land, and hills intermixed. Solar energy, water resources, and good soil are well provided, and the population is dense. South-West China (Region VI) includes the Sichuan and the Yunnan-Guizhou Plateaux. Most of the territory is mountainous and hilly, and the topography quite complicated, so that transport is very inconvenient. The natural and socio-economic conditions are less favourable than for the other two regions in southern China, and agricultural productivity is comparatively low. The Sichuan Basin, however, is an exception and forms an area comparable with the other regions of southern China. South China (Region VII) is in the tropics and subtropics and is warm all the year round. Plains are intermixed with hills and the land has long been intensively cultivated. The region is densely populated and productivity is high. This is the only place where tropical crops are produced in China.

These, then, are the seven regions in eastern China, and the two regions in western China, which, together with the Region of Sea Fishery, make up the 10 agricultural regions of China.

A comparison of these 10 regions with the natural geographic regions described in Chapter 1 shows, not surprisingly, that they are closely related. Quite a few of these regions are very similar, but there are a few differences in the boundaries. The Inner Mongolia and Areas Along the Great Wall Region falls within the eastern part of China in the above analysis, while division in physical terms places it in the western part of China. It is, in fact, a transitional region as previously mentioned. If it is included in the western part of China, instead of the eastern, the relationship between the agricultural and physical regions is as shown in Table 4.1.

The agricultural regions still represent large areas, and of necessity still encompass quite wide variations in agricultural practice. These variations

TABLE 4.1   The regionalization of the ten agricultural regions

Zoning at the 1st level	Zoning at the 2nd level	Ten agricultural regions zoned at the 3rd level		Corresponding to the regions zoned by physical geography
Eastern part	North	I	North-East China	Eastern Monsoon Region
		III	North China Plain	
		IV	Loess Plateau	
	South	V	Middle and Lower Yangtze River Valley	
		VI	South-West China	
		VII	South China	
Western part	North	II	Inner Mongolia and Areas Along the Great Wall	Xinjiang-Inner Mongolia Arid Region
		VIII	Gansu and Xinjiang	
	South	IX	Qinghai and Tibet	Qinghai-Tibetan Plateau Region
–	–	X	Region of Sea Fishery	—

are more closely defined by the sub-regions into which the main regions are divided, but due to the general nature of this book, the sub-regions will only be treated briefly. The tenth agricultural region, concerned only with the marine fishery, has already been discussed in Chapter 3, and will not be discussed further in this chapter.

Region I: North-East China

The North-East China Region is the northernmost region in the eastern part of China and includes the whole of Heilongjiang and Jilin provinces and a part of Liaoning province, excluding Chaoyang prefecture. The region occupies a total area of 0.953×10^6 km². It was known as Manchuria until the twentieth century, as it was the home of the Manchu. Since then it has been known as The Three Eastern Provinces, or simply as the North-East, though the land that the Chinese refer to as The Three Eastern Provinces, or the North-East, may have a western boundary further west than that described here.

In the north-west of the region lies the Da Hinggan Ling Mountain Range, and in the south-east the Changbai Shan Mountain Range, but the rest of the region consists mainly of gentle hills and plains of arable land. The region is well served by rivers: the Heilong Jiang River flows through the north of the region and is navigable, as is the Songhua Jiang River which flows through the middle area. The Liao He River flows through the south-west of the region, and the Yalu Jiang River flows through the south-east. Soils in the region are mainly black earths and chernozems with high organic matter content and a thick humus layer. They have a high natural fertility and are well suited for cultivation. The mountains provide good forestry areas: the overall forest coverage is 30 per cent (35 per cent in Heilongjiang province). Some 30 per cent of the total forestry area of the country and 33 per cent of the timber reserve is found in this region. Here, too, 60 per cent of the country's tusser silk is produced.

The annual rainfall in the region ranges from 450 mm in the west to 700 mm in the east. The winter is long and cold, and it leaves only a short frost-free period of 80–120 days in the north and 140-180 days in the south. The cumulative temperature is also low. The annual cumulative mean temperature for days with mean temperatures of at least 10°C is less than 3000 °C in most of this region; in the north it is only 1300–2000 °C. This means that no crops can over-winter in the field, only one crop per year can be grown, and only spring wheat, not winter wheat, can be grown, except in a small area in southernmost Liaoning province. Fortunately, though, the summer is warm and wet, and growth is good during this season.

North-East China is the home of the Manchu people, and people of the Han nationality were prohibited from migrating into the region during the earlier part of the Qing (Manchu) dynasty. This prohibition was gradually relaxed and Han people moved into the region again in the nineteenth century, and in greater numbers after the overthrow of the Qing dynasty in 1911; however the bitterly cold winters and hard living conditions still limit the population, which is the lowest in the eastern part of China. From the 1950s onwards the government has promoted large-scale cultivation of the land which formerly was lying waste in the region, and some 30 per cent of the present cultivated area has been brought into cultivation as a result of this activity. Today there is 0.3 ha of cultivated land per head of the agricultural population, which is much more than in other regions, and there is more arable waste land available which can be brought into cultivation than in any other part of the country.

There are 17×10^6 ha of cultivated land in the North-East China Region, representing 16.75 per cent of the national total. Most of the cultivated land is devoted to 'grain' crops, and the main crops, in decreasing order of area sown, are maize, soya bean, wheat, millet, sorghum, and rice. Maize accounts for 35 per cent of the sown area and 45 per cent of the 'grain' production in the region — about a quarter of all the maize produced in the country comes from this region. In addition, about half of the country's soya bean, 45 per cent of the sorghum and millet, and 92 per cent of the flax is produced here, and 65 per cent of the area sown to sugar beet is also found here.

Rice is grown in the east and south-east of the region, where almost one million people of Han Chinese or of Korean descent live — their expertise in growing rice in this cold zone is highly spoken of. Nearly one million hectares of paddy rice are grown, amounting to 7 per cent of the sown area of all 'grain' crops in the region. The production from this irrigated rice (see Chapter 3) is 14 per cent of the total production of the 'grain' crops, a reflection of the greater yield of rice compared with the other 'grain' crops. In addition, the quality of the rice produced in this region is better than that of the rice produced in most other parts of the country. Although the yields of the crops other than rice are not high, the annual 'grain' production of more than 500 kg per head of the agricultural population is still the highest in the country. This is due mainly to the comparatively large area of cultivated land per head of the agricultural population, and to the emphasis on growing 'grain' crops. More than 15 per cent of the cultivated land is managed by state-run farms and this, together with the high per caput production, makes the region a producer of commodity 'grain'. The northern part of the region is an area of concentrated potato production and the main area for producing seed potatoes.

In general, cropping in the North-East China Region is extensive. Little manure or fertilizer is applied, there is a low level of management, and the yields are also comparatively low. This reflects the short history of cropping in the region, and the relatively large area of land per unit of labour available — 1.16 ha of cultivated land per unit of agricultural labour force (adult male equivalent) in the whole region, and 1.98 ha in Heilongjiang province. Because manure or fertilizer has rarely been applied, the organic content of the soils has been decreasing, and the humus layer is becoming thinner. This is a matter that must be considered by all concerned with farming in this area in the future. Only 12.4 per cent of the cultivated land is irrigated, a much lower proportion than in the other regions except for the Qinghai and Tibet Region. Another point worth noting is that the cropping pattern in the mid-part of the region is not very sensible. Here, much land is being used to grow maize, a crop requiring a high level of thermal energy and a long growing period, while crops requiring less thermal energy, such as millet and soya bean, are being planted less. This type of cropping pattern is very sensitive to periods of cold weather in the summer or autumn, and as a result the output is unreliable. From 1949 to 1976, in the province of Heilongjiang alone, there were 15 lean years due to low temperature damage of varying severity. In addition, there are droughts and sandstorms in the mid-western part of the region, floods and waterlogging on the flat and/or depressed land, and soil erosion on the mountain slopes.

Timber production in North-East China yields more than half of the country's total output, but long-term deforestation, coupled with a lack of new sapling production, has led to a constant reduction in forest area and timber output over a number of years. The situation has been further aggravated by the absence of a sound fire control system. Further exhaustion of the forest reserves in this region will be inevitable unless effective measures are taken, and if such measures are not taken this region will cease to be an important producer of forest products.

The account given above is a description of the general picture throughout the North-East China Region. There are, however, differences between areas within the region, and these can be considered within the context of the sub-regions shown in Fig. 4.2. The sub-regions of the North-East China Region may be described as follows.

Sub-region I₁: The Forestry Zone of the Hinggan Ling Mountains

This is mainly a mountainous zone in the northernmost part of China. The climate is cold, with 80–120 frost-free days per year, and it is only suitable for a few early maturing crops. Cropping is therefore limited, but it is the most important forestry area in China.

Sub-region I₂: The Agricultural Zone of the Songhua Jiang-Nun Jiang and the San Jiang Plains

This is mainly an area of flat and rolling land of high fertility. The frost-free period is 110–165 days per year, and the annual cumulative mean temperature for days with mean temperatures of 10°C or higher is 2200–3000°C per

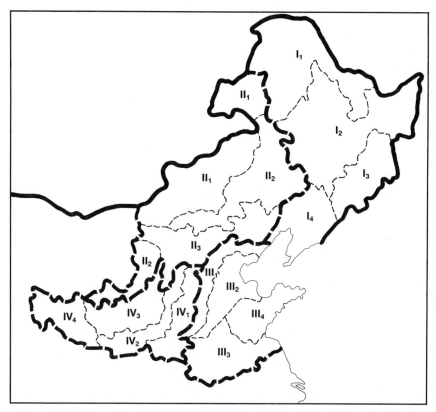

FIG. 4.2   The northern agricultural regions and sub-regions of eastern China.

year. This sub-region forms the main agricultural base of North-East China, and an important national 'grain' producing area; some arable wasteland is still available for future development.

Sub-region I₃: The Forestry-Agricultural Zone of the Changbai Shan Mountainous Area

This sub-region is located in the south-east of North-East China and more than 90 per cent of the area is hilly and mountainous. It is another important forestry area, and some four-fifths of the country's ginseng and pilos antler are also produced here, together with other medicinal materials.

Sub-region I₄: The Agricultural–Forestry Zone of the Liaoning Plain and Hilly Area

This sub-region is in the south of North-East China. It is warmer than the sub-regions to the north, has a longer history of cropping, and the farming is more skilful. However, the population is more dense, and there is only 0.17 ha of cultivated land available per head of the agricultural population. Cropping is the main agricultural activity, but there is also considerable forestry, animal husbandry, and sideline production as well. Fruits, tusser silk, peanuts, and various special items are produced in addition to the main output of 'grain', forest products, and animal products.

Region II: Inner Mongolia and Areas Along the Great Wall

This region is situated in the mid-eastern section along the northern border of China (see Fig. 4.1), and it covers mainly the Inner Mongolia Autonomous Region, though small areas in Chaoyang prefecture in Liaoning province, and in Hebei province, Beijing Municipality, Shanxi province, Shaanxi province, and Ningxia Autonomous Region are also included. The region extends more or less along the line of the Great Wall and includes some areas to the south of the wall. The total area of the region is 0.801×10^6 km^2: 8.34 per cent of the area of China. However, only 3 per cent of the agricultural population of the country lives here, so the population density is less than in other eastern regions.

The northern part of the region is mostly a plateau about 1000 m high. It is rolling land with a few low hills, and forms a basin of internal drainage where only a few rivers are found. In the south of the region there is a small mountainous and hilly area, though the land is lower here than in the north. Many stream beds wind through the mountainous gullies but water only flows along them for a short period after rain. The climate is transitional from the humid and semi-humid climate of the Eastern Monsoon Region to the arid climate of the Xinjiang–Inner Mongolia Arid Region. The annual rainfall decreases from south to north and from east to west, so that the highest rainfall of about 450 mm is in the south-east of the region, and the lowest, at a little over 200 mm, is in the north and west. There is a high annual variation in the rainfall, and strong winds frequently occur in the spring. As a result, drought is common in the region.

Temperatures are low, and winters are long and bitterly cold because of the high latitude and altitude. Snow is not unusual, even in May, and it can appear as early as September. The frost-free period is only 100–150 days, and the annual cumulative mean temperature for days with mean temperatures of 10°C or higher is 2000–3000°C, which means that only one crop per year can be grown. Only cold tolerant crops with a short growing period are suitable, so winter wheat, for example, is not grown. The natural vegetation from the prairie in the east, through the steppe in the middle, to the desert steppe in the west, is grassland, and hardly any woods can be seen. The Hulun Buir Meng Prairie in the north-east of the region has the most luxuriant grassland in China, and can produce 3–4 tonnes/ha of fresh herbage per year. However, the herbage yield in other areas is not high, and in the desert steppes in the west of the region it is very low.

In the past, most of the region north of the Great Wall was inhabited by Mongols, who lived almost exclusively by livestock grazing; south of the Great Wall, the Han, who lived on cropping, were to be found. However, since the nineteenth century the Han have moved into the central part of the region, and the present situation is that the north of the region is still grazing land, the central part is partly grazing and partly cropping, and the south is a cropping area. The Han still engage in cropping, and the Mongols in livestock raising — each national group follows its own calling.

There are at present 7.8×10^6 ha of cultivated land in the region, 99 per cent of which is not irrigated. Most of this cultivated land is used to grow crops of high drought tolerance, such as spring wheat, maize, sorghum, millet, broom corn millet, naked oats, and potatoes. Small areas are also sown to sugar beet, and to cold-tolerant oil crops such as linseed, spring rape, and sunflower. Cultivation is extensive, fertilizers are rarely applied, and management of the crops is comparatively careless. In a few places people still follow the system of abandoning the land after several years of cropping. This reflects the larger area of cultivated land available per head of the agricultural (including pastoral) population of more than 0.3 ha, and the short history of cropping in the region. Also, the low and unreliable rainfall, together with the high winds in the spring, make crop yields very unreliable. In 1978, the yield of 'grain' crops was 1.4 tonnes/ha/year, the lowest of the 10 agricultural regions. In the same year 'grain' production was only 341.5 kg per head of the agricultural population, a little more than that of the Qinghai and Tibet Region, but lower than for the remaining agricultural regions.

Livestock grazing is the main activity in most of the region, and, as mentioned previously, the Hulun Buir Meng Prairie in the north-east has the most luxuriant grassland in China. In the remainder of the region the grassland is not as good, but it still provides comparatively good grazing for China. Livestock raising in the region is basically the exclusive business of the Mongols, who follow a nomadic system of husbandry. In recent years compounded-feedstuff supply centres have been built in a few places where the water supply and soil conditions are favourable, with the aim of releasing the herdsman from their complete dependence on nature. The livestock are mainly sheep, goats, cattle, and horses, the latter two being used as draught animals.

One of the serious problems of this region is that there is an over-emphasis on cropping while not enough emphasis is given to animal husbandry. For several decades the tendency to regard the grassland as waste land, and to subject it to numerous indiscriminate campaigns to transform it into cultivated land, has often resulted in a production increase in 'grain' for only a few years, and the eventual promotion of the desertification process. The fact is, there is only a limited water supply which can be used for irrigation, and the climate is dry and windy — it is very easy to induce wind erosion. The conditions are not suited to cultivation, and the efforts given to expanding the area of cultivated land in the past few decades have resulted in the reduction of the grassland area and the development of desertification, while the total area of cultivated land has not been increased.

Another problem that should be mentioned is the excessive development of stock raising. According to the estimate in *A Comprehensive Agricultural Regionalization of China*, there were 3.2 ha of grassland for each sheep unit in 1949, and this had dropped to a mere 1.3 ha by 1978. If the flocks and herds are too large, and insufficient grass is available, then over-grazing will occur, leading to a loss of grass growth through excessive treading and eating, and a decline in herbage quality. The animals always eat the palatable grasses first, and leave the unpalatable untouched, enabling easier regrowth of the unpalatable species. Both these problems should be kept in mind by people involved with the development of this region.

As mentioned above, this region is a transitional region which can be grouped with the regions of either the eastern part or the western part of China. The region itself can be subdivided into three sub-regions as follows, which depict this transition successively (see Fig. 4.2).

Sub-region II$_1$: The Grazing Zone of Northern Inner Mongolia

The inhabitants of the north and north-western parts of the region are almost entirely engaged in grazing, and the population density is quite low, at only two persons per km². It is a zone which obviously has the characteristics of western China.

Sub-region II$_2$: The Grazing–Agricultural Zone of Central–Southern Inner Mongolia

This sub-region is much more densely populated than the first, with about 50 persons per km², and the people here are engaged either in cropping or in stock raising, which reflects the duality of eastern and western China. The Mongols are engaged in grazing while the Han are engaged in cropping. People of the two nationalities have their own separate life-styles, but their communities may be distributed in a mixed pattern, though that does not mean that grazing and cropping are being carried out together in the one unit. It is easy to understand that this is why there are always contradictions between the Mongol and the Han peoples. The Han always want to expand their cultivated land, which damages the grassland; while the herds of the Mongols often walk over the cropping land, which destroys the crops. The outcome of these contradictions is invariably harmful to the grassland. This, too, is a problem which urgently needs a solution.

Sub-region II₃: The Agricultural–Grazing–Forestry Zone of the areas along the Great Wall

This sub-region is located in the southern part of the region. The population density here is more that 60 persons per km², and most of the people are Han. Cropping is the predominant activity although some grazing is also carried on, and the sub-region shows some of the characteristics of the North-East China Region. However, the rainfall is less than in North-East China, only 25 per cent of the cultivated land can be irrigated, drought is common, and crop yields in most places are rather low. Furthermore, this is a mountainous area and there is serious soil erosion on the mountain slopes; high salinity is a further problem in the basins among the hills. Farming in this sub-region is difficult, and the people here have a hard life. It is one of the poorer areas of China.

Region III: North China Plain

This region is located in the north of the eastern part of China (see Fig. 4.1), and covers an area of 0.444×10^6 km². It is sometimes called the Huang He–Huai He–Hai He River Valleys. In fact it includes not only the North China Plain itself, but also the hilly land of Shandong Peninsula.

In the west, the region includes a few slopes at the foot of the adjacent mountains, and in the east there is the hilly land of the Shandong Peninsula. Otherwise most of the region is an alluvial flood-plain with the Hai He River flowing eastwards to the sea through the northern part, the Huang He River through the middle areas, and the Huai He River through the southern part. Another smaller river, the Luan He River, flows through the northern part of the region in a southerly direction to the sea. Most of the land of this region is formed of deep alluvial deposits from these rivers, and particularly from the Huang He River (the Yellow River). The whole region declines gently from west to east, so that it is more or less flat, with three-quarters of the land below 100 m. A vast sheet of plain is a rare formation in China and this is the largest in the country.

The climate is warm–temperate, with cold winters and scorching summers. The annual cumulative mean temperature for days with mean temperatures of at least 10°C is 4000–5000°C, the annual frost-free period is 175–220 days, and two crops a year or three crops in every two years can be harvested. Nearly all types of crops, including the thermophilic crops such as cotton and peanut, can be grown in this region. The annual rainfall is 500–800 mm. Although this is not low, the variability is high, so the rainfall is not particularly reliable. Moreover, two-thirds of the annual rainfall comes within the three summer months, and there is a shortage of water in the winter and spring. A drought in spring and waterlogging in autumn is not unusual. There is a considerable amount of saline land in the depressions, and more along the sea coast, which influences the agriculture of the region to a small extent.

Historically the region was the centre of Ancient China's politics, economy, and culture, and it was where countless events of great historical importance took place. Although agriculture here has a very long history, most of the cultivated land having been in use for thousands of years, the farmland is still as productive as ever and shows no sign of exhaustion. There are 22.4×10^6 ha of cultivated land, the largest area for any of the agricultural regions. Similarly, the cultivation index — the ratio of cultivated land to total land area — is more than 50 per cent, and higher than for any other region. However, the agricultural population is also large, at 183×10^6, and is second only to that of the Middle and Lower Yangtze River Valley Region. As a result there is only 0.12 ha of cultivated land available per head of the agricultural population. This is about average for the country as a whole, but it is far less than for the five northern and western regions and only a little above that for the three southern regions.

The main crops of the region are wheat, maize, and cotton. The areas sown to wheat and to cotton are the highest for the regions, and the area sown to maize is second only to that of the North-East China Region. The production of wheat is 40 per cent of the national total, of maize 28 per cent, and of cotton 23 per cent, and the output of these crops is ranked first or second among the regions. The areas sown to peanut, sesame, and tobacco are also ranked first among the regions; the areas sown to millet, sorghum, and soya bean rank second. This region is also the leading producer of temperate fruits such as apples, pears, persimmons, and jujubes (Chinese dates). Forest coverage is only 7–8 per cent — just a little more than for Region II and for the two regions in the western part of China. Animal husbandry is not developed: the animal population amounts to only 1.0 sheep units per head of the agricultural population, a figure just a little higher than that for the Middle and Lower Yangtze Valley.

In this region the system of three crops in every two years is common in the dry farming areas, and the system of two crops per year is popular for irrigated land; intercropping and inlaid cropping are also commonly practised. The main crops grown in the dryland system are spring sown crops such as maize, sorghum, millet, and sweet potato, followed by winter wheat, and then summer sown crops such as maize, sweet potato, millet, and soya bean. With two crops per year the main crops are winter wheat with summer sown crops such as maize (predominantly), sweet potato, and sorghum. Cotton, however, is always sown in the spring, as a single crop per year; and continuous planting of cotton is very popular. The development of irrigation facilities and other technical advances in agriculture have increased the cropping index of this region quite markedly during the last 30 years. It reaches 150 per cent for all of the cultivated land, and 170–190 per cent for the land sown to 'grain', that is, the land is almost never allowed to lie fallow.

There is an abundance of groundwater available, most of it of good quality, and this has provided a good basis for the rapid development of irrigation during the last 30 years: 55.8 per cent of the cultivated land is now irrigated. This is a far higher proportion than in the other three regions in the north, and is similar to the average for the regions in the south. However, in recent years the weather in the north has been rather dry, and the amount of water used in industry, and for domestic use in urban areas, has been increasing rapidly, leading to shortages of both ground and surface water. Water shortages have already become quite serious in some places, and there is great interest in the project under consideration to divert water from the Yangtze River to make up the deficiency of water in the region. This project will be enormous in every respect, but it will have to be undertaken some day if the problem of a shortage of water is not to become increasingly serious.

Too much water can also be a problem. Most of the land in the region is rather flat, and the summer rainfall is highly variable. If heavy rain occurs in a wet year, land lying in depressions easily becomes waterlogged. In east Henan and west Anhui, land is frequently waterlogged, as also is land in central Hebei. The best way to overcome this is to construct a complete land drainage system; but constructing such a system gives rise to many social and economic problems which are very difficult to resolve. Besides, the the rivers in the region, and especially the Yellow River, all flood easily in the summer, and so summer flood control is also important.

The flat terrain makes transportation easy: vehicles, or at least horse-drawn carts, can reach everywhere in the region, except for the hilly areas in the Shandong Peninsula. Animal-drawn carts are still the peasants' most widely used means of transport even though there has been some advance in the use of tractors and even trucks. Water transport is, however, scarce in this region. The Hai He River system is non-navigable, the Yellow River (Huang He) is only partly navigable, and the Huai He River, though navigable, is on the southernmost margin of the region. The north section of the Grand Canal, in the east of the region, was navigable and did transport large quantities of goods and materials, but, unfortunately, the water supplied to the Grand Canal was gradually put to other uses and eventually none was supplied to the canal at all. Since the middle of the 1970s this section of the Grand Canal north of the Yellow River has been a dry river bed, or an historical ruin.

In general, peasants in this region have a rich experience of cropping, but are less meticulous than are their counterparts in southern China. Many places in the region are susceptible to drought or waterlogging, and large areas suffer from high salinity. As a result, crop yields are comparatively low. The yield of cotton, for example, for the whole region in 1978 was only 274.5 kg/ha/year, far lower than the average yield for the country as a whole which was 445.5 kg/ha/year, and less than half of the yield for the Middle and Lower Yangtze River Valley Region which had a yield of 654.8 kg/ha/year. The average yield of all the 'grain' crops in the region in 1978 was 3975 kg/ha/year, which was also far less than for the three agricultural regions in southern China.

The main differences between the four sub-regions within this region may be described as follows.

Sub-region III₁: The Agricultural Zone of the Piedmont Plain of the Yan Shan and Taihang Shan Mountains

This sub-region is located in the north-western part of the region (see Fig. 4.2) and mainly consists of the piedmont diluvial–alluvial fan and piedmont plains where the drainage conditions are good, and there are rich resources of good quality groundwater. These attributes have made this the leading sub-region for irrigation development in northern China. The soils are mostly fertile and are suitable for various crops and fruit trees, and yields are generally high and reliable. In general, this is a comparatively good sub-region for agricultural production within the North China Plain Region.

Sub-region III₂: The Agricultural Zone of the Hebei, Shandong, and Henan Depression Plain

Situated in the north-eastern part of the region, this sub-region has a low, flat topography, and inferior drainage; and high salinity is a common problem. As a result productivity is low and is badly in need of improvement.

Sub-region III₃: The Agricultural Zone of the Huang He and Huai He River Basins

This sub-region is on the border adjacent to southern China. The seasonal distribution of rainfall is more regular than in the other sub-regions, and as a result the spring drought is no longer a severe threat. In addition, the land is flat and the soil fertile, so the sub-region is well suited to cropping. It is an important base for the production of many 'grain' and industrial crops, including wheat, sesame, flue-cured tobacco, soya bean, and peanut, which are produced in substantial quantities in relation to the total national output. However, cultivation in this sub-region is not yet meticulous enough, crop yields are not high, and further improvement is needed before the full potential of the sub-region can be realized.

Sub-region III₄: The Agricultural Zone of the Shandong Hilly Area

This area has less cultivated land per caput and more intensively managed agriculture than in any other area in northern China. Consequently, the yield of 'grain' is the highest, the diversified management is the best, and the living standard of the local people is the highest in northern China. (In 1978 there was 0.1 ha of cultivated land per head of the agricultural population, and the yield of the 'grain' crops was 5138 kg/ha/year). This is also the most important peanut producing area in the country, containing 27 per cent of the total area sown to peanuts, producing 32 per cent of total peanut production, and accounting for 80-90 per cent of peanut exports from the whole country. It is also one of the most important fruit producing areas: the apple of Yantai, the pear of Laiyang, and the grape of Dazeshan are all very famous for their quality as well as quantity. This is also the second biggest production area for flue-cured tobacco, and the output of tusser silk is second only to that of the province of Liaoning.

Region IV: Loess Plateau

The Loess Plateau Region is situated to the west of the North China Plain Region and to the south of the Inner Mongolia and Areas Along the Great Wall Region (see Fig. 4.1). It includes the physically defined Loess Plateau, described in Chapter 1, and has an area of 0.406×10^6 km².

As a whole, this region is a plateau, but there are some river valleys in the plateau and some mountainous areas in the south of the region. Apart from the river valleys, most of the region is at an altitude of 1000–1500 m above sea level, but the landscape of this plateau is quite different from that of the Inner Mongolia Plateau. The latter is a uniform plateau of considerable height, and the relative undulations over the ground surface are small. On the Loess Plateau, although the overall altitude is high, and certain peaks are very high, the ground surface is deeply eroded and far from flat.

The Loess Plateau Region derives its name from the widely distributed loess layer on the ground surface in most parts of the region. The loess layer in the west of this region is about 200 m thick, and a decrease in thickness is observed from west to east. The thickness of the loess layer over most of the plateau is several tens of meters, but it is only a few meters in the eastern border area. However, some 30 per cent of the ground surface in the region has the bedrock exposed by erosion of the surface layer, giving a rocky mountainous landscape.

It is generally accepted that the loess was transported from the west by the wind, but whether or not there was subsequent transportation due to the action of flowing water is still being debated. One of the characteristics of loess is that it has a high homogeneity of particle size, with a silt content of over 60 per cent. Other characteristics are a low cohesion, resulting in looseness, high porosity, and a very high susceptibility to water erosion. The soil erosion of the Loess Plateau is very serious. At present 90 per cent of the area is affected by soil erosion, and half of this erosion is very serious indeed. Across the plateau the soil erosion reaches as much as 5000–10 000 tonnes of soil/km^2/year or even more. As a result, the whole Loess Plateau is densely cut with countless gullies and ravines and appears as a criss-cross network of millions of blocks. The gullies and ravines can be as deep as 150 m, and they occupy an area of 40–50 per cent of the total area. Such a landscape not only has a dreadful appearance, but also gives rise to great difficulties for cultivation and transport, and leads to an extreme shortage of water over the plateau.

The climate in the region is mild and temperate, slightly cooler than in the North China Plain Region, and with more features of a continental climate. The annual cumulative mean temperature for days with mean temperatures of at least 10°C is 3000–4000°C per year; and the frost-free period is 120–250 days per year. Three crops can be grown in every two years over most of the region, and two crops a year in some places. The annual rainfall is 400–600 mm: this may not seem low, but the variability is quite high, and a large proportion of the annual rainfall may come in the form of storms, which mean a low water availability. The result of this unreliable rainfall pattern is a frequent occurrence of drought, particularly in spring.

Historically, the south-east of this region was the cradle of the ancient civilization of China and contained many of the capital cities such as those of the Zhou dynasty, the Qin dynasty, and the Han dynasty. The now world-famous funerary terracotta warriors of Qin Shi Huangdi were also unearthed here. There is a long history of cropping in this part of the region, and in the north and west of the region there was once a considerable area devoted to forestry and army horse raising, both of which have since vanished. Today there is little natural vegetation or forest, and nearly the whole region is taken up with crop cultivation. However, the crop yields are much higher in the south-east than in the north-west, and, indeed, the differences between the two parts are extraordinary.

There are 10×10^6 ha of cultivated land in the region, and an agricultural population of 55×10^6. This gives an average of 0.18 ha of cultivated land per caput, but the actual figure could be higher than this, especially in the areas of the loessial gullies and ravines. Most of the cultivated land lies on slopes with inclines of 10–35°, and the plots of land are very small. Flat valley-floor plains, and flat lands on the top of the plateau, are comparatively large in area, but make up a mere 10 per cent of the total cultivated land. Ninety four per cent of the total sown area grows 'grain' crops, 30–50 per cent of which is wheat. Maize is the second most important crop, occupying 25–30 per cent of the 'grain'-crop area. Irrigation is only available for 25.4 per cent of the total cultivated area — this is distributed mainly over the Guanzhong Plain in the south of Shaanxi province, and the Fen He River Valley in Shanxi province. Here the most fertile farm land in the region is to be found, and wheat and cotton are mostly grown. In the other parts of the region dry farming is practised, and 'grain' crops such as maize, millet, and broom corn millet are grown. The yields are very low, at around one tonne/ha/year, and even lower in some places. Animal husbandry is also under-developed and yields about 15–20 per cent of the gross agricultural production. Because of the poor transport there are few other economic activities and the living standard of the people is quite low.

Sub-region IV₁: The Agricultural–Forestry–Pastoral Zone of the East Shanxi and West Henan Hilly Areas

This sub-region is in the eastern part of the region (see Fig. 4.2) and consists of highlands and mountains 1000–1500 m above sea level. There are quite a number of small valleys and basins throughout the sub-region, with rivers flowing through them, and the annual rainfall is 550–700 mm, which makes the sub-region semi-humid. Agriculture is well developed because of the good soils and adequate water supply, and the sub-region is an important area for maize and millet production. The area of cultivated land available per person is low, and cultivation here is reasonably meticulous; the cropping index averages 131 per cent, and the yield of 'grain' crops 3240 kg/ha/year. Forestry and animal husbandry tend to be better than in other parts of the region.

Sub-region IV₂: The Agricultural Zone of the Fen He and Wei He River Valleys

This sub-region is in the southern part of the region. These river valleys were both very important in ancient times, especially the Wei He Valley, which is also called the Guanzhong Plain. In general, the sub-region has a smooth terrain, and the farmland extends without interruption into a vast stretch of land, fertile and free from erosion. Well developed irrigation facilities support a high level of agriculture. The average yield of 'grain' crops has been 3–4 tonnes/ha/year, and that of cotton 380–750 kg/ha/year depending on location, with a fairly stable output from year to year. Wheat production in the sub-region amounts to three-quarters of the joint output of the two provinces of Shanxi and Shaanxi, 'grain'-crop production to more than half, and cotton production to as much as 97 per cent.

Sub-region IV₃: The Pastoral–Forestry–Agricultural Zone of the Shanxi, Shaanxi, and Gansu Loessial Hills and Gullies

This sub-region lies in the northern part of the region and is an impoverished and backward area where soil erosion is serious, crop yields are low, and the rural people have a very hard life. The major problem is the soil erosion which leaves the remaining soil infertile and unproductive. Extensive cropping yields only a scanty harvest, and, in turn, causes further erosion, and so a vicious cycle continues. This sub-region is the main source of the huge amount of silt which enters the Yellow River, and the problem is indeed a grave one.

Sub-region IV₄: The Agricultural–Pastoral Zone of the Central Gansu and Eastern Qinghai Hilly Areas

This sub-region is in the west of the region, and has a high terrain with most of the land at 2000–2500 m above sea level. The annual rainfall is 250–500 mm; the mean July temperature 20°C; the annual cumulative mean temperature for days with mean temperatures of at least 10°C is 2000–3000°C; and the annual frost-free period is 100–150 days. Only one crop per year is grown and the yields are rather low. The sub-region is a transitional area between Region IV in the eastern part of China, and the Qinghai and Tibet Region in the western part of China.

The general features of southern China compared with those of northern China

The climate of southern China is warm and humid, the annual rainfall exceeds 1000 mm except in areas adjacent to North China, and the variability of the rainfall is lower than in northern China. The seasonal distribution of the rainfall is comparatively even, and so there is no marked division between the dry and the rainy seasons as there is in northern China. Droughts do occur in southern China, and they do affect agricultural production, but they normally occur in the summer, whereas droughts in northern China always occur in the spring. Normally the rainfall in summer is sufficient, but sometimes the time interval between two falls of rain may be a little too long, and a short drought is experienced. This temporary shortage of water is made more apparent because the temperatures are high. It is also the period of rapid growth for the rice, and as the evapotranspiration rate is high the crop is sensitive to a lack of water. Since crops other than rice are generally not threatened by drought, all of the irrigation schemes in southern China are built for the purpose of expanding the area planted to rice. If other crops are irrigated it is just incidental to the main purpose. As a whole, southern China is rich in water resources and the area does not suffer the problems of water shortage that are seen in northern China.

The temperature in southern China is, of course, higher than that in the north and thus allows the cultivation of plants such as tea, citrus fruits, and various tropical trees. The leaves of winter wheat do not die off as they do in the north; instead, the plants keep growing, even if only very slowly. The people who live in southern China do not have heating facilities, as the people in northern China do, and the higher temperatures encourage an entirely different building style. The houses in northern China are all of a closed type, while the houses in southern China, or at least the old fashioned ones, are of an open type which cannot be heated to keep the rooms warm. It is quite common in the winter for people living in southern China to go out onto the southern side of their building to be in the sunshine to get warm — the temperature indoors is lower than that outdoors in the sun.

In southern China there are many mountains and hills but few plains, none of which are large. This contrasts with the large plains, such as the North China Plain, to be found in northern China. There are, however, many stretches of water in southern China. The drainage systems, including the natural streams and rivers, and the artificially built canals, are numerous and perform well, particularly in the river valleys, basins, and deltas, where they extend in all directions to form complete networks of waterways.

All over southern China yellow earths, lateritic soils, and laterites are to be found which are high in sesquioxides, clayey, and of low natural fertility. However, thousands of years of cultivation have changed the properties of these soils so much that they are considered by Chinese soil scientists to be developed agricultural soils based on lateritic soils or laterites, and no longer the original soils. This is particularly so for the soils which have been used to grow rice for a long period. The Chinese soil scientists classify this type of soil as a special great soil group and call it a 'paddy soil'. This, they consider, can be formed from any kind of parent material or great soil group, but only if it has been used to grow rice for a sufficiently long time. There are none of these high sesquioxide soils in northern China, but a great variety of great-soil-group soils is to be found there.

With the warm and humid atmosphere, the fields in southern China are nearly always green, in contrast to northern China where for several months each year the scene is greyish yellow and barren. For the same reason it is easier for a southerner to find food; a smaller area of cultivated land can feed a person in southern China because the cropping index is higher and there is a higher yield per unit area per year. This is one of the reasons, perhaps the most important reason, for the very high population density in southern China. If the rather mountainous terrain of the south is taken into consideration, the difference in population densities between the south and the north is even more evident. If the warm and humid conditions exemplify the favourable natural factors in southern China, then the high population density and the rugged terrain characterize the social and historical background.

Thanks to the favourable natural conditions and because of the large population, the agriculture in southern China is typified by rice planting, a high cropping index, and a high yield. The peasants here place special stress on rice, so their land is invariably converted to rice paddies wherever conditions permit — the yield of rice is higher than that for other crops. Most of the land in southern China will bear two crops of rice per year, and still have another crop of wheat or rape or green manure in the winter. In such a cropping system the yield is 7–10 tonnes of 'grain'/ha/year, or even more. In some places the peasants follow a system of one crop of rice followed by one winter crop giving two crops per year, and this yields 6–8 tonnes of 'grain'/ha annually. This latter system requires less labour and has its advantages. However, the winter crops in the systems, including wheat and barley, are not considered to be important by the peasants in southern China, and their yields are not very high. As mentioned previously, this area is, nevertheless, the main source of commodity wheat in the country. A similar situation arises with animal husbandry, since, although it is a minor activity, the output is higher than for other regions, and the commodity rate for animal and poultry products is also high.

Not only is more 'grain' produced in southern China than in northern China, but there is also a higher yield of cotton per unit area. There are many other crops which yield more in the south than in the north: this reflects not only the environmental differences, but also the more meticulous cultivation practices of the peasants in the south. This does not suggest any differences in the natural abilities of the people in the south and the north, rather it is due to the fact that there is a very high density of people and only limited cultivated land available in the south and the peasants there have to strive hard for their livelihood.

The following examples may perhaps help to illustrate this situation. In southern China the earthen partition ridges between the paddy fields are as narrow as 30 cm, but, even so, a line of broad beans is usually planted on either side of them. Walking paths through the fields are often 40–50 cm wide, and sideline crops are planted on both sides of them. Even the roads linking villages are hardly more than 1 m wide. The pathways to walk on, the lodgings to live in, the edges of the village, the land around the house, the courtyard in it, all are narrow. These are the results of a large population and limited arable land; every inch of the land must be made use of. In contrast, the pathways and courtyards in the north are much more spacious. In the eyes of a southerner this seems just extravagant.

Correspondingly, the southern peasants usually travel on foot in the countryside and carry a load on their shoulder, although occasionally they may use a wheelbarrow: this is because of the narrow pathway. In northern

China routes are wide enough to allow for carts (man or animal drawn) to pass. And, of course, people in the south who live within reach of waterways invariably use boats for transport, an option which is rarely available in the north.

Region V: Middle and Lower Yangtze River Valley

This region lies to the south of the North China Plain Region (see Fig. 4.1) and its northern boundary is the famous Qin Ling Ridge–Huai He River line. To the region's west lies the South-West China Region which is actually a part of southern east China, and the boundary between the two regions is the eastern border of the Yunnan–Guizhou Plateau and its extension. To the south of the region lies the South China Region with the Nan Ling Range as the boundary, and to the east the Pacific Ocean. The area of the region is 0.969×10^6 km^2, about one-tenth of the area of the country.

Apart from a plain of moderate size, most of the region is hilly, with small valleys separating the low hills. Plains and river basins account for only one-quarter of the area. A dense network of rivers and streams runs across the plains, together with a large number of lakes: the fresh water in the region accounts for half the area of fresh water in the country.

The climate over the whole region is warm and humid, and the annual cumulative mean temperature for days with mean temperatures of at least 10°C is 4500–6500°C. The frost-free period is 210–300 days per year. Two crops of rice per year can be grown in most parts of the region, as can perennial, subtropical plants such as tung oil trees, tea bushes, citrus trees, Chinese fir, and bamboo. The annual rainfall is 800–2000 mm, with no distinct cut-off between the dry and rainy seasons. In the east of the region typhoons, bringing heavy falls of rain, are common in summer.

As an area of hilly land and extensive water, the region has only 19×10^6 ha of cultivated land, but an agricultural population of 233×10^6 people; hence the area of cultivated land per person is only 0.08 ha. Such a dense population with limited cultivated land leads to intensive cultivation with a high cropping index, and a high yield per unit area per year. The majority of the cultivated land, 72.3 per cent, is in rice paddies. More than half of this grows two crops of rice a year along with winter crop such as wheat, rape, or barley, The remainder of this area grows a crop of rice and then a winter crop in succession. On the dryland two crops a year are grown: a summer crop such as maize, cotton, or sweet potato, and then a winter crop such as wheat, barley, Chinese milk vetch, or broad beans.

The cropping index for the whole region was 216.7 per cent in 1978, the mean yield for the 'grain' crops 6540 kg/ha/year, and the yield of cotton 655 kg/ha/year — the highest figures for any of the regions. The importance of this region in the agriculture of China is apparent when it is considered that 19 per cent of the country's cultivated land is in the region, but this produces 57 per cent of the country's rice, and 33 per cent of the total 'grain'; 43 per cent of the area sown to rape is in the region, and 39 per cent of the area planted to cotton from which is produced 57 per cent of the country's cotton. Some 70 per cent of the country's tea is produced here, 48 per cent of the silkworm cocoons, 75 per cent of the oil-tea camellia seed, and substantial proportions of the country's jute, bamboo, fir timber, and subtropical fruits. This is also where 60 per cent of the freshwater aquatic produce comes from, and where some 31 per cent of the country's pigs are kept.

The region can be subdivided into six sub-regions, as shown in Fig. 4.3.

Sub-region V_1: The Agricultural–Animal Husbandry–Fishery Zone of the Lower Yangtze River Valley Plain and Hilly Area

This sub-region is situated in the north-eastern corner of the region, and it is the most densely populated, economically developed, and agriculturally advanced area in the whole of China. Not only is Shanghai, the most powerful economic centre with the largest population, located in this region, but also such time-honoured cities as Nanjing and Hangzhou, and the

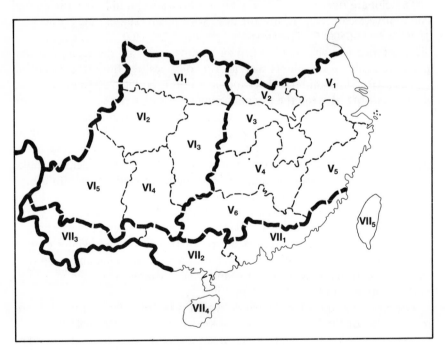

FIG. 4.3   The southern agricultural regions and sub-regions of eastern China.

thriving and developing medium-sized cities of Ningbo, Wuxi, Changzhou, Nantong, and Yangzhou.

The area of the sub-region is $0.15 \times 10^6 \, km^2$, or 1.6 per cent of the country's total, but the output is 11 per cent of the national total of 'grain', 31 per cent of cotton, 38 per cent of silkworm cocoons, and 20 per cent of jute, and 11 per cent of the nation's pigs are to be found there. In general, the living standards and the cultural attainments of the people who live in this sub-region are far ahead of those found in many other parts of China.

Sub-region V_2: The Agricultural–Forestry Zone of the Henan, Hubei, and Anhui Low Hill and Plain Area

This sub-region is situated in the north-west of the region, and it shows some transitional characteristics, as it borders the North China Plain Region in the north. Although paddy cultivation forms the basis of the agriculture, compared with the region as a whole the agriculture here is comparatively backward.

Sub-region V_3: The Agricultural–Fishery Zone of the Middle Yangtze River Valley Plain Area

Situated to the south of the previous sub-region, this sub-region is flat and has an abundance of lakes, including Poyang Hu and Dongting Hu. The whole Jiang Han Plain was formed from depositions in ancient waters, and this origin makes the area very suitable for rice growing and freshwater aquatic production; pig and poultry production are also popular. There is, however, a drainage problem because the land is so low lying.

Sub-region V_4: The Agricultural–Forestry Zone of the Hilly and Mountainous Area South of the Yangtze River; Sub-region V_5: The Forestry-Agricultural Zone of the Zhejiang and Fujian Hilly and Mountainous Area; Sub-region V_6: The Forestry-Agricultural Zone of the Nan Ling Hilly and Mountainous Area

Although there are differences between these three sub-regions, they are all, in general, hilly and mountainous. Both agriculture and forestry are important in this area, but they are managed at a lower level than Sub-regions V_1 and V_3 above. Sub-region V_5 is a coastal zone where the population is large and there is insufficient cultivated land available, and it is from here that in the past many people have gone abroad to make a living or settle permanently. Although emigration is not specifically an agricultural matter, it is of much importance to the rural economy and rural living.

Region VI: South-West China

This region is in the south-western part of eastern China (see Fig. 4.1); it is bounded in the north by the Qin Ling Mountain Range and in the south by a line running roughly between the Tropic of Cancer and the 24th parallel. It differs from Region V in that the mountains and plateaux of this region are much higher than the hills and plains of Region V. The area of Region VI is 1.008×10^6 km^2.

There are many high mountains in the region. In the south is the Yunnan–Guizhou Plateau at an altitude of 1000–2000 m, and in the north there are many mountains and the famous basin, the Chengdu Plain, which lies in the middle of the mountains and extends for 7500 km^2. Most of the northern area is at an altitude of 500–2500 m. Throughout the region small valleys and basins are spread through the mountains and steep ridges.

Altogether, 71.7 per cent of the region is mountainous, 13.5 per cent is hilly, 9.9 per cent is formed into plateaux, and 4.9 per cent is formed into plains (Division of Economic Geography, *The General Agricultural Geography of China*, 1980). Because of the mountainous nature of the land and the consequent difficulties of cultivation, the cultivation index is only 12.3 per cent, and there is only 0.08 ha of cultivated land per head of the agricultural population. Much of the cultivated land is on mountain slopes and is devoted to dryland farming; only 44 per cent is used for rice paddies. On the cultivated mountain slopes the soil tends to be shallow and of low fertility, and erosion is serious as there is a lack of soil conservation measures.

The climate of the region is similar to that of Region V, except that the summer temperatures on the Yunnan–Guizhou Plateau are lower because of its high altitude. As a result neither cotton nor two crops of rice a year can be grown on this plateau. Parts of the region are alpine, and in these areas the frost-free period is so short that only one crop a year can be grown. Furthermore, the northern and central parts of the region are often cloudy, foggy, and rainy; the period of sunshine here is the shortest in the country, and so these areas are unfavourable for photophilic crops such as cotton.

Historically, the northern part of the region was more or less subordinated to the dynastic rulers whose sphere of influence was centred in northern China. The remainder of the region had been considered barbaric, and the guiding principles of the feudal rulers had been more to maintain relations with, and moderate the activities of, the local inhabitants than to subordinate them. A rapid growth in population has only been seen here in the last two centuries. Even today, the region as a whole is still a place where many minority nationalities of various ethnic origins live in compact communities, the largest number of minority groups being found in Yunnan province.

The main crops of the region are rice, wheat, maize, and rape. Although the area of the rice paddies is only 44 per cent of the cultivated land, the output of rice accounts for 57 per cent of the region's 'grain' production. However, the rice paddies are dispersed in the mountains, where irrigation is difficult and the rains in spring are uncertain, and in winter a large number of the rice paddies are used for the storage of water. This kind of rice paddy is called winter paddy, and it differs from the normal rice paddy in that it is not used to grow a winter crop; it is also called a 'look at Heaven paddy', which means rain-fed.

The total 'grain' production of the region amounts to 16.5 per cent of the national total, and this is the second highest for any region. Rice production is 21 per cent of the national total, wheat 18 per cent, maize 17 per cent, and sweet potato 21 per cent. Some 70 per cent of the region's rice is produced in the Sichuan Basin. The area planted to oil-seed rape is 29 per cent of the national total and second only to that planted in Region V. Cotton, sugar cane, and ramie are also produced in amounts that are of national importance.

The cropping index is comparatively low, at 180 per cent for the region as a whole, and only 150 per cent for the Yunnan–Guizhou Plateau. The yield of 'grain' crops is not high, at 4.5 tonnes/ha/year on average for the region, and both the cropping index and the 'grain' yield figures are unusual for southern China. On the other hand, the cured tobacco from Yunnan and Guizhou is renowned for both its quality and yield. Tea from Yunnan and Guizhou, and the citrus fruits, cane sugar, and silk from Sichuan are all important in China. Some 80–90 per cent of the country's tung oil and raw lacquer are produced in this region, and 60 per cent of the oil seed from the Chinese tallow tree. Even the numbers of cattle and pigs are the highest for any region, and the numbers of sheep and goats the second highest.

The weakness of the agriculture, as well as of the rural economy in this region, lies in the remoteness from the sea and the inconvenience of transportation. Navigable watercourses are scarce and there are few railway lines; the high mountains make the expansion of navigation and rail transport laborious. Highways are common but most of them meander through the mountains, making travelling expensive in terms of fuel and time. Air transport is also restricted because of the difficulties of finding landing fields in the mountains, and because of the cloudy weather which can delay flights. These transport problems drasticaly curtail economic development in the region.

The people in the north of the region have a high standard of living, but not so the people in the south, while life for the minority nationalities in the remote and backward areas is even harder.

The region may be subdivided into five sub-regions as follows (see Fig. 4.3).

Sub-region VI₁: The Forestry–Agricultural Zone of the Qin Ling and Daba Shan Mountainous Area

This is an area where the natural conditions are complex and mixed. It is adaptable to forestry, agriculture, and sideline production. The northern boundary of the sub-region is the well known Qin Ling Mountain Range, which is steep and majestic, 400–500 km long from west to east, and 120–180 km wide. The highest peak is 3767 m above sea level. Connecting with the Min Shan Mountain Range and the Daba Shan Mountain Range in the south-west, the Qin Ling Mountain Range forms a complete barricade for South-West China, barring the cold air current from the north in winter, and keeping this sub-region warmer in winter than the Yangtze-Huai He Plain which lies on the same latitude (the monthly mean temperature for January is always above 0°C). The climate on the northern slopes of the Qin Ling Mountain Range is quite different from that on the southern slopes, although the distance between them is small.

Sub-region VI₂: The Agricultural–Forestry Zone of the Sichuan Basin

This sub-region has the best natural conditions, it is the most densely populated, has the highest level of agricultural production, and the highest living standards, of all the sub-regions in Region VI. The 'grain', oilseed rape, mulberry silkworm cocoons, citrus fruits, cotton, jute, tobacco, tung oil, medicinal herbs, and pork produced here are all important at the national level. Sichuan has for long been looked upon as the 'Land of Abundance', and it is not an undeserved reputation.

Sub-region VI₃: The Forestry–Agricultural Zone of the Sichuan, Hubei, Hunan, and Guizhou Borders Mountainous Area; Sub-region VI₄: The Forestry–Agricultural–Animal Husbandry Zone of the Guizhou–Guangxi Plateau Mountainous Area; Sub-region VI₅: The Agricultural–Forestry–Animal Husbandry Zone of the Sichuan–Yunnan Plateau Mountainous Area

The agriculture in these three sub-regions is not particularly good. The natural conditions are poorer than in the other two sub-regions, and the people here lead a comparatively hard life.

Region VII: South China

This region is the southernmost agricultural region of China. It is a tropical and subtropical area and is bounded in the north by the Nan Ling Mountain Range and its extension (see Fig. 4.1). The area of the region is 0.496×10^6 km^2, approximately the same size as the North China Plain Region and the Loess Plateau Region, and much smaller than the other six regions.

The most significant natural features of the region are its high temperature and heavy rainfall, which favour the growth of tropical plants. This is the main reason why this area is designated as a separate region. The monthly mean temperature of the coldest month is above 12°C; the mean annual minimum temperature over a number of years is not below 0°C; the annual cumulative mean temperature for days with mean temperatures of at least 10°C is 6500–9500°C; the annual frost-free period is 300–365 days; and much of the region is free from frost, ice and snow. Plants grow quite rapidly all year round, giving an evergreen appearance to the vegetation. On the coastal deltas and plains in the east of the region three 'grain' crops can be grown per year, and in some places even four crops per year. Eight to ten rounds of vegetables can be grown in a year, 7–8 rounds of silkworm cocoons can be harvested in a year, freshwater fish farms can produce 3–4 catches a year, and tea can be picked 3–4 times a year, all of which makes this region quite different from the other regions. And not only is there this difference in quantity, but there is also a much larger variety of tropical fruits and tropical industrial crops grown in the region. However, some tropical plants, such as rubber trees, are still at risk of damage from low temperatures, and plantations of trees and plants of this type are never wholly successful.

The annual rainfall in most of the region is 1500–2000 mm, although it can be much higher in some places. The highest recorded annual rainfall, of 6489 mm at Huoshaoliao in Taiwan, is the record for China. The seasonal distribution of the rainfall is not even: in winter and spring it is rather low, and crops can easily suffer from drought during these seasons; in summer and autumn torrential rains and typhoons frequently occur, and also harm agriculture.

Hilly and mountainous terrain dominates the region and covers 90 per cent of the total area; river deltas and small plains occupy the remaining 10 per cent. Most of the region is bordered in the south by the sea, and it includes two of China's largest islands: Taiwan and Hainan Dao. The long coast line not only provides the region with a prosperous fishery, but also furnishes a base for overseas communications and a gateway for many people who are making a living abroad: a large number of Chinese emigrants can trace their origin to this region. Because of the steep and rugged terrain only 11.6 per cent of the region is arable, and the over-population means that there is only a very small area of 0.07 ha of cultivated land available per head of the agricultural population. Both of these figures are the lowest for any of the regions. In the coastal deltas and the small plains the area of cultivated land per caput falls even lower, to 0.03–0.05 ha. Economic activity is not limited to the seas and the arable land, however; in the mountainous areas forestry is a major industry, with some 30 per cent of the region covered by forests. This is the highest proportion for any region, and Region VII is one of the three main forestry areas in the country.

The agriculture of the region is centred on rice growing. About 70 per cent of the cultivated land is devoted to rice paddies, and the planted area of rice is 70 per cent of the total area planted to 'grain' crops. The output of rice is 84 per cent of the total output of 'grain' in the region. It is therefore apparent that crops other than rice occupy only a small area. Nevertheless, the planted area of sugar cane in this region accounts for 62 per cent of the total area planted to this crop in the country, and the yields are also the highest in the country at 45–60 tonnes/ha/year in general, with a maximum figure of 150 tonnes/ha/year. Peanut is another important crop and 30 per cent of the sown area in the country is sown in this region; second only to that sown in the North China Plain Region. The system of cropping followed in the region is one of three crops per year, the cropping index is 201.4 per cent, and the average yield of 'grain' crops is 5.3 tonnes/ha/year. Only Region V has a higher cropping index and yield.

The region is the main growing area for China's tropical and subtropical fruits — such as banana, pineapple, litchi, longan (or dragon eye), orange, and tangerine; and for tropical industrial plants — suc' as rubber tree, *Agave* sp., oil palm, coffee, cocoa, lemon grass, pepper, and cinchona. The area planted to these crops is not large, and the output is similarly limited. Rubber is the most important of them, because there is a heavy demand for rubber from industry. In the past hardly any rubber trees were planted; but during the last 30 years many rubber plantations of considerable size have been established on Hainan Dao Island in the south of the region, and at Xishuang Banna in the west.

Region VII may be subdivided into five sub-regions (see Fig. 4.3).

Sub-region VII₁: The Agricultural–Forestry–Fishery Zone of the South Fujian and Central Guangdong Area

This sub-region is famous for the density of its population on the small arable area, for the intensity of the cultivation, for the highly developed rural economy in all its sectors, and for the high living standards of the people. Its location at the 'South Entrance' to China, and close proximity to such international ports as Hong Kong and Macao, makes this sub-region a thoroughfare for incoming and outgoing travellers, and imports and exports. It has a long-established foreign trade, and in the past a large number of emigrants have embarked on their voyage from here. In recent years special economic areas (for the introduction of foreign capital and technology) have been opened up in the region at such places as Shenzhen, Zhuhai, Swatow, and Amoy, making the sub-region a rather unusual area.

Sub-region VII₂: The Agricultural–Forestry Zone of West Guangdong and South Guangxi; Sub-region VII₃: The Agricultural–Forestry Zone of South Yunnan; Sub-region VII₄: The Agricultural–Forestry Zone of Hainan Dao Island, Leizhou Peninsula, and the South China Sea Islands

These three sub-regions are similar except that Sub-region VII_3 is an area where several of the minority nationalities live, particularly the Dai; and Sub-region VII_4, which includes Hainan Dao Island, has the richest resources of water and solar energy. There are development schemes in both of these sub-regions at present, and there are further areas with potential for development within these two sub-regions.

Sub-region VII₅: The Agricultural–Forestry Zone of Taiwan

Detailed information about this sub-region is omitted, except that its area is included within that of the South China Region. The failure to include information about Taiwan in this book will be understood.

Region VIII: Gansu and Xinjiang

This region is the northern part of western China (see Fig. 4.1). To the north-east and north-west it has a long border with the People's Republic of Mongolia and the USSR, and to the west a shorter border with Afghanistan and Kashmir. Internally, it is bounded in the south by the Kunlun Shan and Qilian Shan Mountain Ranges, and in the east by approximately the 250 mm isohyet. The region consists of the whole of Xinjiang Uygur Autonomous Region, the western part of the Inner Mongolia Autonomous Region, and the northern parts of Gansu province and the Ningxia Hui Autonomous Region. The total area is 2.254×10^6 km², or 23.48 per cent of the territory of China; thus the region is similar in size to the Qinghai and Tibet Region, and is far larger than the other regions. It is an arid and sparsely populated area.

The land of the region lies on the Mongolia–Xinjiang Plateau at 700–1500 m above sea level and is without much undulation, though sloping slightly from south to north. The whole region is surrounded by high mountains, except in the north-east where a flat open border area lies between China and the People's Republic of Mongolia. The Tian Shan

Mountain Range extends deep into the central west of the region; and there are also some low lying basins including the well-known Turfan Depression (or Turpan Pendi Basin), which lies below sea level.

Throughout the region the climate is arid. Where the annual rainfall is less than 250 mm, the aridity, K, is generally greater than 2.5. However, over more than half of the region the annual rainfall is less than 100 mm, and here the aridity, K, is greater than 4.0. Under such conditions cropping is impossible without irrigation. Fortunately, many high mountains, which enjoy a high rainfall, surround the region or are to be found in it. For example, the annual rainfall in the Tian Shan and Qilian Shan Mountains is 400–600 mm. There are also permanent snow packs and glaciers in the high alpine areas, over 3500 m above sea level: these partially melt and the water flows out in the summer. This is the main source of irrigation water for the local oasis agriculture. The only exception is the Ili He River Valley between the Altai Shan Mountains and the Tian Shan Mountains in the north-west corner of the region, where the north-west side is wide open and exposed to the air currents from the Arctic Sea. The annual rainfall there is about 400 mm, and crops can be grown without irrigation. This is only a small fraction of the region, and only 2 per cent of the whole region is cultivated. More than 95 per cent of the area is desert, semi-desert, and mountains. This includes such well known deserts as the Taklimakan Shamo Desert in south Xinjiang, the Gurbantünggüt Shamo Desert in north Xinjiang, the Badain Jaran Shamo Desert in west Gansu, the Tennger Shamo Desert in north-central Gansu and Inner Mongolia, and the Mu Us Shamo Desert also in Inner Mongolia. The deserts cover a total of 0.60×10^6 km^2, or one-quarter of the region.

The winters in Region VIII are bitterly cold, and the summers rather warm. The annual cumulative mean temperature for days with mean temperatures of at least 10°C is 2600–4300°C, and the annual frost-free period is 120–180 days. However, the great majority of the days in the year are clear and sunny.

This is a region where many people of the minority nationalities live. The Uygur and the Hui mainly devote themselves to cropping, while the Mongols, the Kazak, the Kirgiz, the Tajik, and the Yugur engage in livestock grazing. Since the 1950s the Chinese government has set up a number of state-run farms to develop cropping and animal husbandry in the region, and at the same time quite a large number of the Han nationality have immigrated into the region and expanded the area of cultivated land quite substantially. There are now 15×10^6 people in the agricultural (including pastoral) population, which is 1.85 per cent of the national agricultural population. The area of cultivated land is 4.5×10^6 ha and this gives a figure of 0.3 ha of cultivated land per person.

Some 86 per cent of the cultivated land is irrigated: rain-fed cropping is only possible on a few alluvial fans or plains to be found at the foot of various high mountains, and the yields here may be only a fraction of those obtained from the irrigated land. The cultivated land in the region is found in the piedmont plains where water is available for irrigation, and in the eastern part of the region in the vicinity of the Great Bend of the Yellow River. In both of these cases the distribution of the cultivated land is restricted, and the areas are limited by the availability of irrigation water. Because of the scarcity of water only a few types of crops are grown. In most cases, wheat occupies more than half of the sown area, and much of this is spring wheat. The next most important crop is maize, and long-staple cotton and sugar beet are also of importance. Normally, only one crop a year is grown. Cultivation tends to be extensive in form, and there is a widespread and serious problem with salinity. These factors together result in low yields — for 'grain' crops the average is only 2.3 tonnes/ha/year. However, as there is quite a large area of cultivated land per person, 'grain' production in the region is still more than sufficient for local requirements.

To date this region has generally been regarded by the public as a pastoral region; the area used for herding and grazing is many times larger than that used for cropping. However, the animal husbandry of the region is carried out under the conditions of the dry climate, bitterly cold winter, and very poor pasture of the region. A transhumant system of husbandry is followed in which the grasslands in the high mountains are grazed in summer, the desert grasslands of the piedmont basins are grazed in winter (these lie to the leeward of the mountains and are generally sunny in winter), and the savannah grasslands of the expansive piedmont plains and desert, between the above two areas, are grazed in spring and autumn. The livestock raised in this region show the features of desert animals, the main types being fur-bearing sheep, wool-bearing goats, and camels. It is a primitive type of husbandry, of low productivity, yet it does effectively utilize the very limited resources available. There are, however, many problems. The availability of grasslands across the seasons is uneven, and there are gaps in the supply; herbage productivity is very low, there are no back-up resources to help overcome natural adversities, and it is easy for disagreements to arise between people tending herds and flocks and those growing crops. It is worth noting that 90 per cent of the well-known Xinjiang donkeys are raised in south Xinjiang. They are easily reared in this system, and can be used for rural transport over short distances. Although the number of livestock slaughtered in the region is 4.6 sheep units per head of the agricultural population per year, the total production in terms of animals slaughtered is still only 6.2 per cent of national production.

There are very few large stretches of forest in the region, except for those in the north-west mountainous area, and water bodies suitable for aquatic production are even rarer.

The region may be subdivided into three sub-regions (see Fig. 4.4).

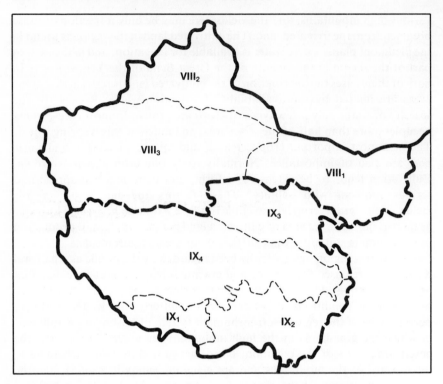

FIG. 4.4 The agricultural regions and sub-regions of western China.

Sub-region VIII₁: The Agricultural–Pastoral Zone of Inner Mongolia, Ningxia, and Gansu

This sub-region includes the irrigated areas of the Great Bend of the Yellow River in Ningxia and Inner Mongolia, and the piedmont lands of the Qilian Shan Mountains in Gansu province which are supplied with water from the snow thaw. These are the main 'grain' producing areas, and a high proportion of the 'grain' is sold on the market rather than being consumed by the producers.

Sub-region VIII₂: The Agricultural–Forestry–Pastoral Zone of Northern Xinjiang

Here the climate is slightly wetter than in other parts of the region, and so the grassland is a little better. In addition there are 1.7×10^6 ha of forest land, which is three-quarters of the forest land of Xinjiang province.

Sub-region VIII₃: The Agricultural–Pastoral Zone of Southern Xinjiang

This large sub-region occupies more than half of Region VIII. The duration of sunshine is 2800–3300 hours annually, the annual cumulative mean temperature for days with mean temperatures of at least 10°C is 3900–4300°C, and the frost-free period is about 200 days per year — thus the hours of sunshine, and the temperature levels, are very suitable for growing upland cotton and long-stapled cotton of medium and late maturing varieties. The limited irrigation areas within the sub-region are also famous for growing fruits and melons; grapes from Turpan (Turfan) and musk-melon from Hami are well known throughout the country.

Region IX: Qinghai and Tibet

This region is situated in the south-west of China, and its territory is identical with that of the Qinghai–Tibetan Plateau Region defined in Chapter 1. It embraces the whole of the Tibet Autonomous Region, the major part of the province of Qinghai, and small parts of the provinces of Gansu, Sichuan, and Yunnan (see Fig. 4.1). The area of the region is 2269×10^6 km², which is 23 per cent of the whole country. The agricultural population of the region, however, is a mere 0.5 per cent of the national total agricultural population, the area of cultivated land 0.7 per cent of the national total, and the 'grain' output 0.4 per cent. It is a high, alpine area with a vast expanse of land and limited population.

   Two-thirds of the region is more than 4500 m above sea level, and the remainder is at an altitude of 3000–4500 m, except for a few river valleys in the south and east which are below 3000 m, and which occupy one-tenth of the region. Where the altitude is higher than 4500 m, the monthly mean temperature of the hottest month is below 10°C, and in some places lower than 6°C; and there is virtually no frost-free period. Cropping is not practicable, and the only way that the land can be used is for it to be grazed by animals tolerant of the cold. Most of the places lower than 4500 m are also unsuitable for cropping, except for the few river valleys in the south and east. However, the duration of sunshine in the region is 2600–3200 hours per year, and the sun shines 60–70 per cent of the time. This is a very good sunshine regimen for crop growth and as a result the yields of the crops which are grown in the region are often very high. Extensive areas of grassland are available for grazing, and in the east and south-east of the region the herbage yields are good. Animal products from this region account for 11 per cent of the national production, and 29.2 sheep units per year are slaughtered per head of the agricultural population. In the south-east and east of the region there are a number of forests, which together make up 23 per cent of the national timber reserve and the second largest forest area in China.

The main domestic animals of the region are the altitude-tolerant yak, the Tibetan sheep, and the Tibetan goat; and the main crops are naked barley, wheat, peas, potatoes, and oil-seed rape, all of which are cold tolerant. The forests consist mainly of spruce and fir. The species common in warmer areas include cattle and pigs amongst the domestic animals, maize and rice amongst the crops, and broad-leaved trees: these can only be reared or grown in the lower-lying river valleys in the south-east of the region.

The region may be subdivided into four sub-regions, as shown in Fig. 4.4.

Sub-region IX₁: The Agricultural–Pastoral Zone of Southern Tibet

The northern part of this sub-region is formed by the valleys of the Yarlung Zangbo Jiang River and its tributaries, where there is a long tradition of cropping on a compact area of cultivated land. The altitude is 3500–4100 m, the topography is slightly sloping, there is a deep soil profile, the annual rainfall is 400 mm, and the annual frost-free period is 120–150 days — the area is, therefore, suitable for the growth of cold tolerant crops.

The 'grain' output of the sub-region accounts for 50 per cent of that of the Tibet Autonomous Region, and for 70 per cent of the commodity 'grain' of the Autonomous Region. In other words it is the granary of the Qinghai and Tibet Region.

The southern part of the sub-region is located on the south Tibet Plateau where cropping intermingles with animal husbandry. In the pastoral area of the sub-region the annual rainfall is 250–350 mm, and the pasture grows fairly well. In addition, the intermingling of cropping and animal husbandry makes it easier to supply supplementary feed to the animals when needed.

Sub-region IX₂: The Forestry–Agricultural–Pastoral Zone of Sichuan and Tibet

In this sub-region the north lies at a higher altitude than the south, and the topography is quite complicated. There are many high mountains and deep canyons, with differences in elevation of up to 1000–2000 m or more. In some places a whole series of natural geographic belts, consisting of highland tropics, subtropics, warm-temperate zones, alpine cold zones, and up to alpine névé zones, all occur within 100 km. The sub-region in general has an abundant rainfall, and nearly all the forests of Region IX are to be found in this sub-region.

Sub-region IX$_3$: The Pastoral–Agricultural Zone of Qinghai and Gansu

Located in the north-east of the Qinghai–Tibetan Plateau, this sub-region lies beyond the Tibet Autonomous Region, and mainly in Qinghai province, but also in a small part of Gansu province. The climate here is mixed: arid in the west, and somewhat humid in the east, and cropping and animal husbandry co-exist. The cropping area is divided into two sections: the river valleys of the Yellow River and its tributaries in the east of the sub-region, and the area bordering the Qaidam Pendi Basin. The latter area was developed in the 1950s and 1960s, and the thawing snow water from the mountains to the north of the Basin is used for irrigation.

Sub-region IX$_4$: The Alpine Pastoral Zone of Qinghai and Tibet

This sub-region includes more than half of Region IX and comprises the main part of the Qinghai–Tibetan Plateau. Because of the high altitude, cold, and dryness, crop cultivation is not possible in the region, and trees do not grow here either. The only way to use the land here is to graze it with animals such as yaks and Tibetan goats which are tolerant of the cold. However, the herbage that grows here is dwarfed and growth is very slow so that the stocking capacity of the sub-region is low.

References

The following texts are in Chinese:

Division of Economic Geography, Institute of Geography, Chinese Academy of Sciences (1980). *The General Agricultural Geography of China.* Science Press, Beijing.

Division of Economic Geography, Institute of Geography, Chinese Academy of Sciences (1983). *The Production Pattern of Chinese Agriculture*. Agricultural Publishing House, Beijing.

Editorial Group of the Plant Industry Regionalization of China (1984). *The Plant Industry Regionalization of China*. Agricultural Publishing House, Beijing.

National Committee for Agricultural Regionalization (1981). *A Comprehensive Agricultural Regionalization of China*. Agricultural Publishing House, Beijing.

Song Jiatai (ed) (1986). *The Economic Geography of China*. The Central Broadcasting and Television University Publishing House, Beijing.

5

Infrastructure and agricultural inputs

Irrigation, flood control, and drainage

Wang Weixin

Irrigation

Irrigation systems have a long history in China. The earliest recorded irrigation system was the Que Bei (Peony Pond), a reservoir project in the years 598–591 BC, to the south of the Huai He River in the present day province of Anhui. The renowned Dujiangyan irrigation scheme built in 256–251 BC has been in service for more than 22 centuries and is now watering farming areas of 587 000 ha in Sichuan province. However it was not until about the third century BC that there was an increase in irrigation projects. Before the eighth century AD such schemes were mainly constructed in North China, and thereafter mainly in southern China.

Since 1949 the government has devoted much attention to agriculture, and irrigation systems have been constructed on a large scale. The size of the irrigated area has been more than doubled and in 1986 it amounted to 44.23×10^6 ha or 46 per cent of the country's cultivated land. According to information from the Food and Agriculture Organization of the United Nations, China has the largest irrigated area in the world.

Because irrigation under different conditions has different features, it is better to divide the country into different areas and to describe the irrigation in each area. For this purpose China can be divided into four main areas: southern China, North China, North-East China, and North-West China. There is also a small irrigated area in Tibet. The discussion in this section is based on statistics for which the province or autonomous region has been taken as the basic unit. Therefore, any province or autonomous region which has irrigated areas with differing features can only be treated in terms of the dominant features. Thus, all of the irrigated land in Shaanxi and Henan provinces will be considered as within North China, while all of the irrigated land in Anhui and Jiangsu provinces will be treated as being within southern China. Similarly, all the irrigated areas of Gansu, Ningxia, and Inner Mongolia, are grouped within North China. The result of this is that there are 13 provincial units in the southern China area, 10 in North China, three in North-East China, and two (Xinjiang and Qinghai) in North-West China, and one further unit — Tibet.

In 1984 southern China embraced an irrigated area of 23.04×10^6 ha, or

52 per cent of the country's total irrigated area; for North China the corresponding figures were 16.47×10^6 ha, or 37 per cent; for North-East China 2.03×10^6 ha, or 5 per cent; for North-West China 2.80×10^6 ha, or 6 per cent; and for Tibet 0.11×10^6 ha, or 0.25 per cent. Clearly, southern China and North China are the main irrigation areas.

Southern China, with its reasonably well distributed rainfall of more than 1000 mm/year, needs almost no irrigation for growing dry crops, and so rice dominates the irrigated land there. Dry crops are grown only in winter or in places where rice-growing is unsuitable. There are mountainous areas in southern China where in order to plant rice the ground must be levelled and water storage facilities such as ponds constructed, and these may be so costly that they can neither be afforded by the peasants nor be economically justified. As a result, there is only a limited distribution of rice paddies in the mountains. In contrast, irrigated rice paddies are widespread and very popular on the flat land, and they may cover 90 per cent of the total area in summer. Taking southern China as a whole, about 60 per cent of the cultivated land is irrigated, but about 90 per cent of China's rice growing is undertaken on this irrigated land. Thus the importance of irrigation in this area, not only to the local community but to the country as a whole, is evident.

The situation with the provinces, autonomous region, and municipalities of North China is different. Here, with the lower rainfall, two-thirds of which falls in summer, droughts are not unusual in the spring. There is a popular saying among the peasants that: 'Rain in spring is as valuable as cooking oil. The harvest will be expected if there is a successful sowing.' It means that rains in spring are scarce; timely rain will ensure a successful sowing and then more rain will come and a good harvest can, in general, be expected. There is a need for irrigation, although irrigation does not necessarily make a lot of difference in summer, because usually there is already enough rain during that season. The major crops in the irrigated area in North China are wheat, maize, and cotton — which is obviously different from the situation in southern China. These three crops are the major crops throughout North China, but a higher proportion of the irrigated land is planted to them than of the dry land. The drought tolerant, but less palatable, crops such as millet, sweet potato, and sorghum are planted to a greater extent on the dry land and less on the irrigated land. In general, irrigation increases the yields in North China, especially in dry years; though in years with sufficient rainfall irrigation may help only a little. There is only a limited area of rice paddies in North China, most of which are small and located in places where spring water is available, or where depressions in the ground accumulate water naturally. Again, vegetables in North China are generally irrigated, but the total area is quite small. In all, about 42 per cent of the cultivated land in North China is irrigated.

In the three provinces of North-East China another picture emerges. The cold climate with extended winters does not permit the growth of winter crops or the production of more than one crop a year. It is, however, warm enough to grow rice in summer, and rice brings higher profits than other crops — this is especially true in the east of North-East China, where about half of the irrigated land is devoted to rice. In North-East China as a whole, though, the main crops, in order of magnitude of planted area, are maize, soya bean, wheat, sorghum, millet, and rice takes only sixth place. In 1984 only about 13 per cent of the cultivated area of North-East China was irrigated.

The Xinjiang Autonomous Region and Qinghai province in North-West China belong to the arid region, especially Xinjiang. Except for the Ili He River Valley in north Xinjiang, cropping is impossible in most places without irrigation, and 82 per cent of the cultivated land in this area is irrigated; in the Xinjiang Autonomous Region alone, 91 per cent is irrigated. The main crops in Xinjiang are wheat, maize, and cotton; and the main irrigated crops in Qinghai are wheat, oil crops, and potato. In Qinghai, although the main dryland agricultural area is lower than 3000 m above sea level, the main irrigated area is above 3000 m and this is too high an altitude for maize and cotton to be grown successfully.

A large number of reservoirs and water diversion structures have been built and many tube wells have been installed, in order to supply water for the irrigation schemes mentioned above. The official statistics show that at the end of 1986 there were 82 716 reservoirs, with a total capacity of 4432×10^8 m^3. Among these there were 350 large reservoirs each with a capacity greater than 10^8 m^3, and with a combined total capacity of 3199×10^8 m^3; 2415 medium-sized reservoirs each with a capacity in the range 10^7–10^8 m^3, and with a combined total capacity of 666×10^8 m^3; and 79 951 small reservoirs each with a capacity of less than 10^7 m^3, and with a combined total capacity of 567×10^8 m^3. In 1984, there were some 6 192 000 ponds and dikes with a total volume of 264×10^8 m^3. These water storage facilities supply about one-third of the country's demand for irrigation water. Official publications also show that there were 2 675 000 tube wells in the country in 1984, located almost without exception in North China where they supply a considerable proportion of the water needed for irrigation. The remainder of the irrigation water is supplied from numerous diversion structures and pumping stations. There were 455 616 permanent irrigation and drainage pumping stations in 1986, with a total power of 19 181 500 kW.

Sub-irrigation techniques are little used, and similarly, it is quite rare for pipelines to be used to convey irrigation water. Various measures have been taken in recent years to line canals to reduce seepage, and to facilitate the operation and maintenance of the canal systems, particularly in the areas with a more advanced economy. Sprinkler and drip irrigation were both introduced into China in the mid-1970s, and there are about one million hectares being irrigated through these techniques. However, it is unlikely that there will be a significant development of these two types of irrigation in China as a whole, and particularly not in southern China, though there may be some further progress in the future. Of particular interest in recent years has been the development of powered water-lift irrigation. Water lifting devices were used in China more than 1000 years ago; but these devices were powered by man, animal, windmill, or watermill, and were only capable of a small lift and were few in number. Before the 1950s only a small number of modern irrigation pumps were in use in a few places in coastal southern China, but since the 1960s the area irrigated by water raised by the use of electric or diesel pumps has increased dramatically (see Table 5.1). Naturally, the low cost, quick return, and simple operation of these pumps have ensured that the peasants prefer them. Nevertheless, it was the development of large scale machinery and energy industries in China that made this mechanization of irrigation possible.

In association with the development of irrigation using pumped water, some schemes have been installed where water is pumped up a particularly high lift. Areas irrigated in this way are found mainly on the Loess Plateau in Gansu, Ningxia, Shanxi, and Shaanxi provinces, and particularly in Gansu province. The lift in some of these regions is 300 m or more, and the highest is 745 m. It is obvious that investment in the construction of these irrigation

TABLE 5.1   The use of motor driven irrigation and drainage equipment, 1952–86

Year	Total irrigated area	Area irrigated by motor driven pumps		Number of motors used for irrigation and drainage	Total power rating of these motors
	$(10^3$ ha)	$(10^3$ ha)	(%)	(10^3)	$(10^3$ hp)
1952	19 959	317	1.6		128
1957	27 339	1202	4.4		564
1962	30 545	6065	19.9	367	6147
1965	33 055	8093	24.5	558	9074
1978	44 965	24 895	55.4	5206	65 575
1981	44 574	25 231	56.6	5672	74 983
1984	44 453	25 062	56.4	6159	78 663
1985	44 036	24 629	55.9	6164	78 245
1986	44 226	25 032	56.6	6507	82 174

Note: The motors are mainly either electric or diesel.
Source: Editorial Board of China Agriculture Yearbook (1986). *China Agriculture Yearbook 1985*. Agricultural Publishing House, Beijing. (In Chinese); Editorial Board of China Agriculture Yearbook (1986). *China Agriculture Yearbook 1985 (English Edition)*. Agricultural Publishing House, Beijing; Editorial Board of China Agriculture Yearbook (1987). *China Agriculture Yearbook 1986 (English Edition)*. Agricultural Publishing House, Beijing.

schemes, as well as the costs of operation and maintenance, are very high, and that they have to be subsidized by the government. This is one reason why these schemes were not installed earlier, the government could not afford such subsidies, and also the industrial backup was not available. But the most important reason for increasing the irrigated area by this means in recent years has been the pressure of population. As there is fertile land on the Loess Plateau, and irrigation can double the yield under the semi-arid conditions there, it is perhaps worth developing these schemes even although they seem economically unreasonable in themselves.

It can be seen in Table 5.1 that the total area of irrigated land in China increased substantially between the 1950s and 1970s. However, since the end of the 1970s this area has remained stable and even decreased a little, and although various undesirable factors resulting from the rapid increase can be corrected, it is most unlikely that such a marked increase in the area of irrigated land will be seen again. Nevertheless some small scale developments, at a slower rate, are still possible. The reasons for this are simple. In the southern part of China, planting rice is almost the sole purpose of all irrigation effort. In recent years rice has become the desired crop wherever the land supports rice-growing. But further expansion of rice paddies, though possible in some places, has become more and more difficult. In North China the limited water supply and the increasing expansion of industrialization and urbanization will result in less and less water being made available to agriculture for irrigation. In both these parts of China, too, the available farm land is already densely populated and intensively utilized, and this further adds to the difficulty of increasing the irrigated area. In North-West China the water supply available for either agriculture or industry comes from the rather limited water resources of the mountainous regions, which have been almost fully utilized. It will be quite difficult to enlarge the irrigated area here. The proportion of irrigated land in North-East China is not large, and there is some room left here to expand. However, from the national point of view, expansion of irrigation in this region, if it did occur, would not make much difference to the overall picture. Thus, it seems that the major efforts to be made from now on should be directed towards improving the technology of irrigation so that more economic and efficient use is made of the water resources available, and thereby a more profitable result obtained from the irrigation activities of the nation.

Flood control and drainage

According to the National Bureau of Statistics there are from several million to more than twelve million hectares of cultivated land flooded each year, and up to two-thirds of the flooded area suffers a one-third or greater loss of crop. Flooding, including river overflow and waterlogging, means adversity.

Flood control is a matter of the utmost importance because the flooding of rivers does not mean damage to agriculture alone; metropolitan cities and industrial centres are located in the lower reaches of large rivers and are also at risk. For this reason, 170 000 km of river embankments and more than 1000 large and medium-sized reservoirs have been constructed. Water pumping stations, with a total rating of more than 59×10^6 kW have also been installed for flood and waterlogging control. Most rivers have been harnessed to the extent that the worst flood likely to occur in a 20–30 year period is guarded against, and critically troublesome reaches have been made capable of combating the worst flood likely to be met with in a period of from 40 to 100 years. Even so, exceptional floods may still violate the defences and cause disasters, and it is floods of this type that cause a considerable area to be flooded every year.

The efforts made to contain dyke-bursting floods in the Yellow River are exceptional. The mean annual flow in this river is only 686×10^8 m^3, but it carries 16×10^8 m^3 of silt each year. This invariably results in massive sedimentation, particularly where the river flow is sluggish, and the river bed rapidly becomes shallow with mud. The shallower the bed, the unsteadier the water course, and throughout history the bursting of the dykes on the Yellow River and the subsequent diversion of the river has occurred many times. Dykes have been built on both banks, but as the silt keeps settling and the river bed keeps rising the dykes have to be enlarged and reinforced. For more than 30 years the state has been contributing very large sums of money and much manpower to raising and widening the dykes in three successive campaigns. The river has now been stabilized in its present course, at least since the last back-diversion in 1946.

Looking back at the 20 years from 1919 to 1938, the Yellow River saw 11 peak flows of above 10 000 m^3/s which caused seven dyke bursts. In contrast, the river has seen 12 peak flows above 10 000 m^3/s in the last 40 years and no dyke bursts have occurred, even when a super flood of 22 300 m^3/s attacked Huayuankou (the Garden Gap) in July 1958. The success has therefore been great. None the less, 1000 km of the river bed in the lower reaches is 3–5 m, sometimes 8–9 m, above the ground outside the dykes, and the bed is rising at a rate of 10 cm every year. This suggests that there will be great difficulty with further raising and widening of the dykes. Controlling the river in this manner is by no means a long term strategy, but so far there has not been any better solution to the problem.

Disastrous waterlogging, caused either by excessive rainfall or by river overflow followed by water accumulation in low lying land and depressions, is another aspect of flooding. Drainage is the measure required to deal with this.

Lixiahe, a region of several counties in north Jiangsu, was known for its frequent waterlogging before the 1950s, before the construction of Jiangdu Pumping Station. This is a giant station with an installed capacity of

49 980 kW, and a pumping capacity of 400 m³/s. A complete canal system of drainage/irrigation was also constructed. Since then, large areas of rice have been planted. In the dry seasons Jiangdu Pumping Station pumps water from the Yangtze River into the scheme, and in seasons when water is excessive the station pumps water from the scheme into the Yangtze River. With both drought and waterlogging abated, the region has become a granary.

Struggles against droughts and waterlogging are not always successful, and there are cases of waterlogging in China every year. Today the regions relatively susceptible to waterlogging are the embankment and river network regions in southern China, the depressions in the North China Plain, and the low land in North-East China.

The embankment region is a group of areas reclaimed from riverside or lakeside depressions by building dykes and pumping stations. The ground surface of these areas is generally at the same level as, or at a lower level than, the river or lake surface outside the dyke. The embankment area relies entirely on the dykes for protection against flooding. The dominant crop in this area is rice: there is no lack of water, and there is no energy cost in obtaining the water. However, the excess water has to be pumped out when there is a heavy rainfall, otherwise the crops will be inundated and damaged. If the dykes burst during a peak flow there will be a disaster. This is why flood prevention is vitally important for embankment areas.

In the delta river-network region the land also stands only a little above the water level, and with a spring tide or typhoon rainstorm this region can hardly avoid being flooded. Although there is a certain irrigation/drainage capacity installed in this region, the problem is that the present criterion for waterlogging prevention is not high enough.

The region in the North China Plain, also susceptible to waterlogging, includes three areas of lowland, in northern Anhui, eastern Henan, and central Hebei, respectively. The annual rainfall of 600–800 mm is not very large, and it is particulary low in spring. However, heavy rain is common in the summer and a considerable proportion of the rainfall becomes runoff, accumulating in the depressions and leading to waterlogging. Generally speaking, this is a region which depends on land drainage systems, a quite different situation from that in southern China. There are, in fact, many drainage ditches, but the problem is that they are not good enough and/or they were constructed individually and lack an integrated plan. Most of the areas affected by waterlogging in the North China Plain are also places where a problem of salinity may occur. Improvement of the drainage system will benefit the reclamation of saline land, and raise the yields of crops on these soils.

Like North China, North-East China faces problems of waterlogging due to flooding or heavy rains. Much work remains to be done here, such as the construction of land drainage systems and the advancement of means of harnessing the floods.

References

The following texts are in Chinese:

Editorial Group (1985). *The History of Water Conservancy in China*. Water Resources and Electric Power Press, Beijing.

Institute of Water Conservancy and Hydroelectric Power Research, Chinese Academy of Sciences (1985). *Collected Research Papers, Vol. XXII, Water Resources, Irrigation and Drainage, History of Water Conservancy*. Water Resources and Electric Power Press, Beijing.

Ministry of Water Resources and Electric Power (1984). *Modern Water Conservancy in China*. Water Resources and Electric Power Press, Beijing.

The following text is in English:

Ministry of Water Resources and Electric Power P.R.C. (1987). *Irrigation and Drainage in China*. China Water Resources and Electric Power Press, Beijing.

Agricultural chemicals

Mei Dunli

Chemical fertilizers

Chemical fertilizers were not used in China until they were imported at the beginning of this century, though Chinese peasants have always been highly skilled in the application of manures. In 1909 Imperial Chemical Industries of Great Britain (I C I) first introduced ammonium sulphate to the Chinese market and called it 'feitian fen' (field fertilizer). Despite the remarkable effect on yields, which were much increased, there was little market in China for this fertilizer. Apart from the high cost, quite a number of the Chinese peasants were worried about the temporariness of the quick effect produced, and about a possible long-term, harmful influence on the soil texture. Although there was some progress in sales afterwards, the market was limited to the vicinities of cities, where there were transport facilities. During the forty years from 1909 to 1949 only 3 000 000 tonnes of commercial chemical fertilizer was imported. In 1952, not long after the People's Republic of China was founded, the whole country consumed only 78 000 tonnes of chemical fertilizer (expressed as content of N, P_2O_5, K_2O). This corresponded to an extremely low 0.6 kg of chemical fertilizer per ha sown. In other words, the country was relying on organic manures and was using hardly any chemical fertilizer. More than 30 years later, the situation had changed greatly. According to the official figures, the total consumption of chemical fertilizer (N, P_2O_5, K_2O basis) and the consumption of fertilizer per hectare sown, in the following years, was:

1970 3 512 000 tonnes;   24.5 kg/ha
1975 5 369 000 tonnes;   35.9 kg/ha
1980 12 694 000 tonnes;   86.7 kg/ha
1984 17 398 000 tonnes;   120.0 kg/ha
1985 17 758 000 tonnes;   123.6 kg/ha
1986 19 306 000 tonnes;   133.8 kg/ha

Thus, there has been a very rapid increase in the use of chemical fertilizers in recent years, and organic manure is no longer regarded as the main fertilizer.

The rapid growth in the application of chemical fertilizers is, of course, associated with the rapid growth of China's chemical industry. As can be seen in Table 5.2 the total output of chemical fertilizers in China was very much greater in 1984 than in 1950. Without doubt, the fertilizer producers owe their success to the great importance that the state has given to agriculture, which has enabled the rapid growth of the chemical fertilizer industry. Indeed, in the 31 years from 1952 to 1982, more than half of the capital invested in the chemical industry was devoted to the chemical fertilizer industry, and 80 per cent of this went to nitrogenous fertilizer producers. Nitrogenous fertilizer is what the soil in China requires, and its use by far exceeds that of any other fertilizer (see Table 5.2).

The manufacture of nitrogenous fertilizer began in China in the 1930s, the

TABLE 5.2   The production of chemical fertilizer in China, 1951-84[a]

Year	Total output (10³ tonnes)	Nitrogenous fertilizer (10³ tonnes)	Phosphatic fertilizer (10³ tonnes)	Potassic fertilizer (10³ tonnes)
1950	15	15		
1955	79	78	1	
1960	405	196	193	16
1965	1726	1037	688	1
1970	2435	1523	907	5
1975	5247	3709	1531	7
1980	12 321	9993	2308	20
1981	12 391	9857	2508	26
1982	12 780	10 218	2537	25
1983	13 789	11 094	2666	29
1984	14 603	12 211	2360	32

[a] All quantities are on a pure N, P_2O_5, and K_2O basis.
Source: Editorial Board for the Yearbook of World Chemical Industry, Science and Technology Information Centre, Ministry of Chemical Industries (1986). *Yearbook of World Chemical Industry 1985/86*. China Social Science Publishing House, Beijing. (In Chinese)

only type produced being ammonium sulphate. There were a number of producers, including the well-known Yongli Chemicals. In 1942 the five producers then active turned out a record 226 000 tonnes of commercial ammonium sulphate. Then came the chaos of the war and it was difficult to keep production going. In 1950 the output was only 15 000 tonnes (expressed as N); production was then increased gradually during the 1950s, and rapidly during the 1960s and 1970s (see Table 5.2).

Between 1951 and 1984 there were two major changes. The first came in the 1960s as a result of the government's vigorous advancement in the late 1950s of a nationwide campaign to set up thousands of small-scale chemical fertilizer plants at county level. These plants included some phosphate plants but most of them made nitrogenous fertilizer using the carbonization process developed in China for the synthesis of ammonium bicarbonate. This process is characterized by the shortness of the flow circuit, the simplicity of the equipment, and the ready availability of the raw materials. By the 1960s such plants were rapidly coming into operation, but difficulties also emerged. These small plants, with few exceptions, were confronted with incompetence in both technical and business management. Wastefulness gave rise to high production and operating costs, and the product, NH_4HCO_3, had poor chemical stability. Later, readjustment of the plants was carried out, and hundreds were closed down, either temporarily or permanently. By 1983 there were 1215 plants of this kind still in operation, producing 9 457 000 tonnes of synthetic ammonia, or 30 260 000 tonnes of NH_4HCO_3, amounting to 56.4 per cent and 57.3 per cent of the country's total production respectively.

The second change came in the late 1970s when 13 whole fertilizer plants were introduced from the USA, Japan, France, and the Netherlands, each capable of producing 1000 tonnes of synthetic ammonia, or 1620–1740 tonnes of urea, per day. These plants were soon completed and brought into operation, followed by a number of urea plants designed in China and equipped with Chinese equipment. These developments account for the continuing increase in nitrogenous fertilizer production.

As will be apparent from the above, the major nitrogenous fertilizers in China are NH_4HCO_3 and urea. In 1983 and 1984 the total outputs of nitrogenous fertilizer, and the quantities of the different types produced were:

	1983	1984
Total	11 094 000 tonnes	12 211 000 tonnes
NH_4HCO_3	6 439 000 tonnes	7 241 000 tonnes
Urea	3 359 000 tonnes	3 726 000 tonnes
Other nitrogenous fertilizers	1 296 000 tonnes	1 244 000 tonnes

That is, about 59 per cent of the output is NH_4HCO_3, 30 per cent urea, and 11 per cent various other forms of nitrogenous fertilizer.

The earliest production of phosphatic fertilizer dates from 1942 when the Yudian Phosphate Plant, Kunming, produced calcium superphosphate using phosphate ore as the raw material, just as it was mined from Kunyang. The daily output was one tonne yet the marketability of the product was still poor and the plant closed down within 6 months. Production did not resume until the middle of the 1950s. Later, the output of phosphatic fertilizer grew rapidly with the help of the government. By 1983, the total output had increased to 2 666 000 tonnes composed of 72 per cent superphosphate, 27 per cent calcium magnesium phosphate, and 1 per cent other phosphates. As compared with the total production of nitrogenous fertilizers, the production of phosphatic fertilizers in China needs to be expanded; it also needs to be orientated more towards the production of high-phosphorus containing fertilizers and phosphorus-containing compound fertilizers. The problem is that China lacks high-grade phosphate ores. There is some low-grade ore in the country, but the best mines are found in areas with poor transport, and this is unfavourable for the development of a phosphate fertilizer industry.

The first batches of potassium fertilizer made in China were produced in 1958. In 1957 the brine in Carhan Salt Lake in Qinghai province was found to contain 1–3 per cent KCl, and to represent a commercial reserve of 153 000 000 tonnes of KCl (net KCl, here and following). The Carhan Potassium Fertilizer Plant was soon built, and put into operation the next year, producing 20 000 tonnes of KCl in 1960. The plant was then closed for a long time and there was no development. In 1983 a new potassium chloride plant capable of producing 40 000 tonnes of KCl per year was constructed and put into operation, and another potassium chloride plant, the Qinghai Potassium Fertilizer Plant, with an annual output of 200 000 tonnes, is under construction. A number of small potassium fertilizer plants have also been in operation since the 1960s, but their total output is very small.

Trace element fertilizers have been used in China for more than 20 years. In 1964 molybdenum-enriched fertilizer was made in Jilin, and its application to soya bean crops resulted in excellent increases in yield. This provided an inspiring start to the use of trace elements. In 1983 the output of trace element fertilizers throughout the country exceeded 20 000 tonnes. There were at least 24 different specifications, mainly of Mo, B, Zn, Fe, and Cu-bearing fertilizers, and 3 000 000 ha of farm land were treated with them.

So far the supply of chemical fertilizers has been inadequate to meet the peasants' requirements in China. To make up the gap, huge amounts of fertilizers have to be imported. In the early and mid-1970s, from 5 000 000–6 000 000 tonnes (N, P_2O_5, K_2O basis) were imported from abroad each year. At the end of the 1970s, these imports increased to some 7 000 000 or 8 000 0000 tonnes per year, and in the 1980s imports have exceeded 10 000 000 tonnes per year. Most of the imported fertilizers are nitrogenous, though smaller amounts of phosphatic and potassic fertilizers have also been imported: since the 1970s some 10 per cent have been compound fertilizers.

Although the country's farm production could be increased by importing fertilizers, large sums of hard currency would have to be spent, and this is not a permanent solution in any case. For this reason, the Chinese government is striving to increase further the output of chemical fertilizers. However, it is inevitable that some chemical fertilizers will continue to be imported to help attain a balance between revenue and expenditure in foreign trade.

At present, chemical fertilizer is supplied to the user through the National Supply and Marketing Co-operative which runs a nationwide network for supply and marketing in the cropping areas. The annual or seasonal demand for chemical fertilizer is reported to each higher unit in a step-wise process from village level, to county level, and in some cases to a higher regional level. A quota based on the overall picture is then decided upon at head office and supplied to the users. Chemical fertilizers have always been a commodity in great demand, particularly the types of fertilizers that the peasants trust, such as urea. Because of the demand, the government has made high-quality fertilizer a privileged commodity, sold to encourage the production or sale of certain farm produce for government purchase. For instance, the state has stipulated that people or units that fulfil or exceed the state's purchase quotas for food grains, pork on the hoof, or cotton, be awarded a warranted supply of a certain amount of high-quality fertilizer (together with other bonuses) in proportion to the sale volume. This is an effective policy through which the state encourages production, as is verified by practice.

Pesticides and herbicides

Pesticides derived from minerals and plants have long been used in China, but only in small quantities. Synthetic organic agricultural chemicals were first made in China in 1942 in Chongqing, where research workers at the Central Agricultural Experimental Station succeeded in developing a process for the manufacture of DDT, a small amount of which was prepared for social hygiene needs. Before long, though, production stopped. In 1949 there were just a few workshops in the whole country producing commercial white arsenic, calcium arsenate, lead arsenate, copper sulphate, rotenone, and such like, with a total output of only a few dozen tonnes.

In 1950, the year after the founding of the People's Republic, a plant for making DDT and another for 666 were built, being put into operation the following year. In the subsequent years many plants of this nature were built. Construction of the first organophosphorous insecticide plant was begun in 1956 and it was commissioned in the following year. Thereafter, plants making orgaophosphorous insecticide gradually increased in number. In 1959 the annual output of this type of insecticide totalled 2500 tonnes, and it increased to more than 100 000 tonnes by 1970, making up one-third of the manufactured agricultural chemicals.

Meanwhile the development and production of fumigants, herbicides, plant growth regulators, and germicides were also under way. Since the late 1970s several varieties of pyrethroid have been developed, and prepared on a number of production lines. As the development of other agricultural chemicals proceeded the production of DDT and 666 was reduced. Finally, in 1983 the State Council of the People's Republic demanded that manufacture of DDT and 666 be prohibited throughout the country. Obviously the transfer to new farm chemicals caused a decline in annual output — from 532 000 tonnes (commercial form, corresponding to 193 000 tonnes of active constituents) in 1980 down to 298 500 tonnes (commercial form) in 1984. However, there was a larger gross tonnage of highly effective pesticide in 1984 which corresponded to 181 700 tonnes of active constituents, very close to that in 1980.

Today China produces 117 types of agricultural chemicals including 52 insecticides, 35 germicides, 24 herbicides and plant growth regulators, and six others. Production exceeds 1000 tonnes per annum for 16 of the insecticides, seven of the germicides, and the two fumigants. Of the latter, the main insecticides are dipterex, DDVP, rogor, oxyrogor, malathion, phoxim, methyl 1605, and 1605; the main germicides are benziminazole-44, kitazine, isokitazine, and sodium pentachlorophenolate; and the two fumigants are chloropicrin and aluminium phosphide. Insecticide production once made up 85 per cent of agricultural chemical production, but this fell to 75 per cent by the mid-1980s, while germicide production rose from 4–5 per cent to 17 per cent of the total. Herbicide production has always been small in quantity, making up only 9–15 per cent of the agricultural chemicals. Organophosphates have formed the major group of insecticides and accounted for 72.8 per cent of the total quantity of insecticide produced in China in 1984.

The lack of synthetic agricultural chemicals before the 1950s meant that plant diseases and insect pests often proliferated beyond control, and spread over large areas in plague proportions. A serious plague of locusts attacked Shandong in 1927 and 7 million people were forced to leave their homes. In 1929, another serious plague of locusts attacked south-east Shandong, with huge swarms blocking the sun and sky and burying the railway on landing; they even stopped trains running. At that time the only measure that could be taken was the mobilizing of millions of people to swat the nymphs on their appearance. The insects could be killed to a certain extent by this means, but at the cost of enormous manpower and money; yet if not done, however, there was tremendous loss. Early in the 1950s people still had to resort to this method since the agricultural chemical industry was still in its infancy and only a small amount of insecticide was being produced. As output increased by the year, so the number of people who had to be mobilized to swat nymphs decreased. In 1951, 100 tonnes of insecticide and 190 million man-days of labour were used to control the locusts; in 1952, 2000 tonnes of insecticide and 43 million man-days of labour; in 1953 9000 tonnes of insecticide and 1 million man-days of labour. Since 1956 the locusts have been fully controlled with insecticides and no plague has occurred.

According to an estimate published by the agricultural authorities in 1980, the extent of the damage caused by plant diseases and insect pests was 146 700 000–160 000 000 ha-occasions in an average year (1 ha damaged three times in a year is 3 ha-occasions). Of this 120 000 000 ha-occasions were reclaimed by using agricultural chemicals, and 15 000 000 tonnes of food 'grain', and 6 000 000 piculs (1 picul = 50 kg) of cotton were recovered. However, due to the lack of variety and quantity of agricultural chemicals, there was still an annual loss of 5–10 million tonnes of food 'grain' and 2–3 million piculs of cotton.

At present the main trends in China are toward an increase in herbicides; the development of bionomic agricultural chemicals and natural bioactive substances; the development of new types of agricultural chemicals which will be safer to spray, dust, and sprinkle while still being effective; and the development of agricultural chemical industries in the remote, outlying districts.

Agricultural chemicals, like fertilizers, are supplied to the users through the National Supply and Marketing Cooperative. The specifications and quantities of pesticides and herbicides to be supplied are estimated either on the forecast of invasions of insect pests or by experience. Special requests should be submitted in advance of when the items will be needed, so that the Cooperative will have time to stock them. If a sudden plague of pests occurs, creating an emergency, both the agricultural officials and the Cooperative branches would inform the units higher up and arrangements would be made to bring the necessary chemicals to the rescue. The requirements can usually be supplied, but occasionally the supply fails.

References

The following texts are in Chinese:

Editorial Board, China Today Series (1986). *Chemical Industries in China Today*. China Social Science Publishing House, Beijing.

Editorial Board for the Yearbook of World Chemical Industry, Science and Technology Information Centre, Ministry of Chemical Industries (1986). *Yearbook of World Chemical Industry, 1985/86*. China Social Science Publishing House, Beijing.

Hu Xiaotong (1986). The production and application of farm chemicals in China. Unpublished paper presented to the Sixth International Conference of Farm Chemical Science.

Animal feedstuffs

Yu Yifan

There was no animal feedstuff industry at all in China before the 1960s. Animal husbandry as a whole at that time was lingering in a state of minor household production. Whether on state-run collective farms, or at the peasants' houses, animals were fed on anything available and there were no manufacturers or suppliers of specially prepared feedstuffs. In the 1960s it was realized that the consumption of feedstuffs, taken as a nationwide matter, had been unreasonably wasteful, though it might seem acceptable for individual small units. Both the waste of 'grain' and the poor profitability became more and more evident. Consequently, factories were built to produce mixed feed made from miscellaneous food 'grains' and the by-products from 'grain' processing (such as husks of rice, bran of wheat, and bean cake) after comminution and mixing. A part of the resulting product was even sold to other villages or areas. This was the beginning of China's feedstuff industry. In the mid-1970s, an up-to-date chicken farm was established in Beijing and at the same time a compound feedstuff plant was introduced from Hungary and soon put into operation. This was the beginning of China's compound feedstuff industry. Since then, quite a number of mixed feedstuff and compound feedstuff plants have been built in many places throughout the country. Early in the 1980s plants for making feedstuff additives were also established. In 1984 the total yield of mixed and compounded feedstuffs exceeded 12 000 000 tonnes, and in 1986 it exceeded 15 000 000 tonnes. According to official data published in *Renmin Ribao* (The People's Daily), 10 January 1985, it was planned that the annual output of mixed and compounded feedstuffs would be 50 000 000 tonnes in 1990, and 120 000 000 tones in the year 2000.

Because most of the animals and poultry, and the available feed, including the green/coarse fodder and the feed that can be exploited, are to be found in the agricultural areas, the general policy for the development of feedstuffs is to draw on local resources, to process in the local area, and to meet the local demand. The state, in turn, will devote its efforts to the building of plants for producing feedstuff additives, concentrated feedstuffs, and pre-mixed feedstuffs which involve the use of a high level of technology and a large investment; private ownership or collective-run factories of this nature will be allowed in special circumstances. In 1986, China established a Committee of Standardization of Feedstuff Technology which supervises the setting of official standards for feeding layer chickens, meat chickens, milk cows, and bacon pigs; for compound feedstuffs for pigs and chickens; for 12 mineral additives; and for the 10 vitamin additives, all of which are to be enacted soon. The drawing up of standards for animal husbandry sanitation is also under way.

However, the feedstuff industry in China is still at an early stage of development. Processed feedstuffs are mostly fed to pigs and chickens, and are seldom fed to other livestock and fish. These feedstuffs are mainly supplied in the form of meal, though a little of the output is pelleted. The variety and the quantity of additives is still small, and this deficiency has affected the production and quality of the compound feedstuffs. Furthermore, too much dependence is placed on maize as a raw material, and hardly any emphasis is placed on other possibly exploitable resources. Another big problem is the shortage of protein feedstuffs. In a country with a population of 1 100 000 000 people, raw protein for producing protein feedstuffs must be supplied on a basis of self-reliance. But much of the bean cake, cotton-seed cake, and rape-seed cake produced is used as fertilizer and for the time being it is impossible to transfer this material to use for feedstuffs, quite apart from the problem of removing the toxic component in cotton-seed and rape-seed cake. Quality control is also a big problem.

Since the early 1970s China has been importing fish meal from Peru; and since 1980 has been importing feedstuff additives such as vitamins and amino acids in addition to fish meal. In the last few years such imports were worth several tens of millions of US dollars. It can be expected that in the next few years these imports will increase both in variety and in quantity.

Before the late 1970s the supply of feedstuffs to users was based on a ration system determined by the government. This system was abolished in 1979 when the economic reform policy came into effect, and users can now make their own choice. But the present feedstuff supply system still needs to be perfected in order to meet the users' demands for quality feedstuffs.

Agricultural machinery

Ke Baokang

China has a long agricultural history and Chinese peasants have developed many tools and technologies in the course of the thousands of years of their agricultural production. Early in the Shang kingdom (c.1520–c.1030 BC) the Chinese used oxen to draw ploughs, though the practice was not yet common, and during the early stages of the Warring States period (480–221 BC) they began to use cast iron and developed the iron plough. In the Han dynasty, during 140–86 BC, a high ranking official named Zhao Guo invented the seed drill drawn by one ox and held by a man, which allowed the peasant to fulfil three functions in one step: opening a furrow, dropping the seed, and covering it with soil. It was very much like the modern drill in structure as well as in basic principle (Fig. 5.1). The keel water wheel (or

北耕兼種圖

麥黍粱
具皆用此

種

鐵尖 鐵尖

FIG. 5.1 Northern China seed drill with iron tips, Ming dynasty.
Source: Song Yingxing (1637). *Tian Gong Kai Wu*. From a facsimile of the first edition, in the Oriental Collections, British Library, Ref. 15226 b. 19.

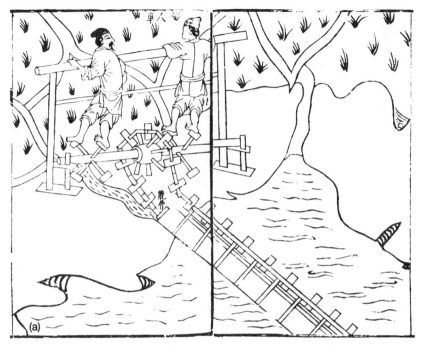

FIG. 5.2(a) Dragon-bone water-lift operated by man power, Ming dynasty.

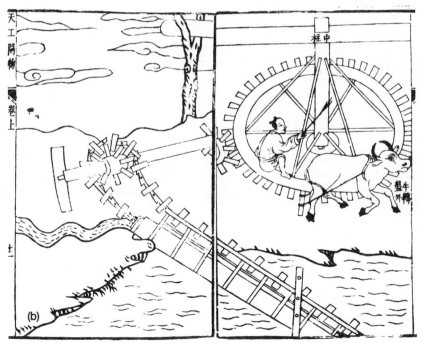

FIG. 5.2(b) Dragon-bone water-lift operated by animal power, Ming dynasty.

FIG. 5.2(c)   Bamboo-pipe water-lift driven by water, Ming dynasty.
Source: Song Yingxing (1637). *Tian Gong Kai Wu*. From a facsimile of the first
edition, in the Oriental Collections, British Library, Ref. 15226 b. 19.

dragon-bone water-lift) invented by Bi Lan in A D 168–189 was widely used in
agriculture until the middle of the twentieth century (Fig. 5.2). There were
also many tools developed for the purposes of harvesting, threshing, and
grain processing very early in Chinese history. However, modern farm
machinery has been developed mainly in Western countries.

The first farm machinery firm established in China was a branch of the US
International Harvester Company, which had moved from Vladivostok to
Harbin in 1922. At that time most of the farm machinery imported was from
that company. After their invasion and occupation of China's three north-
eastern provinces in 1931, the Japanese built many machine manufacturing
plants in Shenyang, Anshan, and Dalian, some of which made farm
machinery, and Chinese businessmen also started to produce some farm
machinery. The largest farm machinery factory in the 1930s in China was
that in Suzhou, which had 120 workers in 1935, and made items such as 4 hp
gasoline engines, 12 and 16 hp diesel engines, water pumps, rice husking
machines, walking ploughs, drills, cultivators, rice threshers, and mulberry
scissors. However, practically all of these farm machinery factories were
closed down after the Anti-Japanese War commenced in 1937.

In December 1943 the Chinese government established the China Farm

Machinery Company, and after the Second World War this company received United Nations relief and carried out the objectives of the UNRRA (United Nations Relief and Rehabilitation Administration), to build factories to manufacture farm machinery in China. At the end of 1946, the first lot of factory equipment and farm machinery (worth $15 million), including tractors, farm machines, water pumps, and implements both manual and animal drawn, was shipped to China from the USA. The Hongjiang Factory (now called the Shanghai Machine Tool Factory) and the Wusong Factory (now called the Shanghai Diesel Engine Factory) were then established in Shanghai, and branch factories were built in Nanjing, Guangzhou, Liuzhou, Changsha, and other places. According to the *China Yearbook*, 1948, volume 2, the monthly production of various farm tools and machines in China was: 1805 units of farm machines (including ginning machines, rice husking machines, rice threshers, etc.), 1495 water pumps, 39 diesel engines (total, 1535 hp), nine gas engines (total, 336 hp), and 95 other engines (total, 375 hp). However, most of these factories closed after 1947 due to the civil war. It was reported that in the middle of 1949, there were only 300–400 tractors (all imported) along with some husking machines, ginning machines, oil presses, and pumps, in use throughout China.

After the establishment of the People's Republic of China in 1949, the new government attached great importance to agricultural production. In addition to instituting land reform and the distribution of land to peasants, the government also paid great attention to increasing the number of traditional farm tools, encouraging the use of modern farm implements through extension, and demonstrating the use of farm machinery. The government supported 50 state-owned factories and 120 private workshops to speed up the production of farm implements. The result of this policy was an increase of 59 million in the number of traditional farm tools available and an increase of 60 000 in the number of modern implements available by the end of 1953, so meeting the demand for agricultural tools in most areas.

As the performance of the modern farm implements was better than that of the traditional tools — they could be used to plough more deeply, harrow more finely, sow more uniformly, harvest more quickly — the peasants were in favour of using them. For example, the new type of walking plough, the three-tined cultivator, the 'emancipation' pipe-and-chain pump, and the rice thresher, were very popular with the Chinese peasants in the 1950s.

As agricultural production developed, the government introduced various tractors and other types of farm machinery from the USSR and other Eastern European countries. At the same time state farms and tractor stations were established in order to demonstrate the superiority of agricultural mechanization to the peasants. In 1953 there were only 11 tractor stations which served the peasants by ploughing and sowing, but this figure had increased to 352 by 1957. They held 7504 tractors of various types, and ploughed 1.836×10^6 ha of land, corresponding to 1.6 per cent of the total cultivated land of China. These farms and tractor stations played an important role in demonstrating the use of farm machinery, in helping to increase agricultural production, and in finding a way of establishing agricultural mechanization in China.

In 1953 the Chinese government began to set up tractor, automobile, combine harvester, and various other farm machinery factories. At the end of 1957, the Chinese factories already had the capacity for large-scale manufacture of multi-furrow ploughs, disk harrows, drills, and combine harvesters. By 1959 they were capable of producing on a large scale various tractors, both wheel and crawler types, various trucks, and cars. Thus, by this time an agricultural machinery industry of considerable scale had been established in China.

Educational and research institutions were also established as the farm machinery industry developed. The Beijing Institute of Agricultural Mechanization (now the Beijing Agricultural Engineering University) was founded in Beijing in 1952, and the Chinese Academy of Agricultural Mechanization Sciences was also set up in Beijing in 1955. By 1959 each province had its own agricultural mechanization research institute, and many agricultural colleges or universities had a department of agricultural mechanization. At the end of the 1960s most counties had their own farm implement research units. Scientific workers of these organizations learnt from imported technologically advanced machines, and at the same time developed new machines to suit the special conditions of China. The rice seedling transplanter and the cultivating boat are two typical examples. The research work on the rice seedling transplanter was started in 1953. In 1956 the first prototype was completed, and then various types of the transplanter appeared. A manual type was put into use in 1965, and a motor-driven transplanter was produced on a large scale in 1967. However, they have never been used widely in China for many reasons, including those of a technical and economic nature, and because of the types of cropping systems. Some of the rice paddies are swampy in origin, and their bottoms are so soft that tractors cannot go into the fields. Chinese technicians therefore developed a cultivating boat which could perform various functions in these swampy fields.

The people's communes appeared in 1958. At that time all of the agricultural cooperative teams were organized into communes, and all of the state-owned tractor stations were handed over to the people's communes which thereafter managed them. Unfortunately, the economic capability of the people's communes was generally very low, and they did not have the technical and administrative abilities to manage large-scale agriculture. The efficiency with which the farm machinery was used was very low, accidents occurred frequently, and the change was a great failure.

Nevertheless, the Chinese government still expended great effort in support of the development of farm mechanization in China. It proposed that Chinese agriculture should be mechanized within 20 years, and in 1969 decided that each county should establish a farm-machinery factory to make and/or repair farm machines. As a result, about 90 per cent of all the counties in the country had their own farm machinery factory by the end of 1970. These factories formed a widespread network so that, in general, farm implements of small and medium size could be manufactured, and tractors repaired, at the local level. The production of Chinese farm machinery, and the level of agricultural mechanization, rose steadily during the period from 1969 to 1979.

In 1979 the total power rating of the machinery in use in the rural areas of China was 182×10^6 hp. This machinery included 667 000 tractors of large and medium size (of 20 hp or higher rating), 1 671 000 small tractors (of less than 20 hp); 1 313 000 farm machines of large and medium sizes, and 97 000 trucks (Table 5.3). The land ploughed with tractor-drawn implements was 42×10^6 ha in 1979 — 42 per cent of the total cultivated area of the country. A total of 10.4 per cent of the sown area of the country was sown with tractor-drawn implements, 2.6 per cent of the sown area of the country was harvested with power driven harvesters, and 0.7 per cent of the rice area of the country was transplanted with rice transplanters (Table 5.4). By this time nearly all the primary machinery demanded for use in the Chinese rural areas, including that used in cropping, forestry, animal husbandry, fishery, and side-line industry, could be made in Chinese factories. In other words, the Chinese farm-machinery industry could now more or less meet the domestic demand for machinery for use in agricultural production.

History had already shown that the people's communes were not good for the Chinese people, and neither was the system of tractor stations managed by the commune cadres. The Chinese government then decided to start economic reforms in 1979, disbanding the communes, changing the collective production system into individual or cooperative production systems, and introducing the contract responsibility system, linking income with output. As these changes brought the peasants into a position of decision making for themselves, their initiative increased steadily, resulting in the rapid development of agricultural production.

Then arose a new phenomenon. Individual peasants, or several families jointly, began to purchase tractors and implements. By 1982 there were one million privately owned and managed tractors; by 1984 this figure had risen to 3.1 million and represented 75 per cent of all the tractors used in agriculture in China; and by 1986, with the continued increase in the number of small and walking tractors in particular, there were 4.7 million tractors owned by peasants — 88 per cent of all the tractors used in agriculture. A similar growth was seen in the private ownership of other items of agricultural machinery and of trucks. By 1982 more than two-thirds of the processing machines for grain, cotton, oil, and feed used in the rural areas were owned by the peasants themselves, and the peasants also owned 89 000

TABLE 5.3   The main items of farm machinery used in China, 1952–86

	1952	1957	1965	1973	1979	1982	1984	1985	1986
Total rated power of all farm machinery (10⁴ hp)	25	165	1494	6503	18 191	22 589	26 557	28 434	31 100
Total power of all tractors (10⁴ hp)	5	63	366	1388	4810	6328	7628	8309	9232
Number of tractors of 20 hp or more (10³)	1	15	73	234	667	821	855	852	865
Total rated power of tractors of 20 hp or more (10⁴ hp)	5	63	363	1062	2861	3604	3702	3730	3804
Number of tractors of less than 20 hp (10³)			4	302	1671	2299	3298	3824	4528
Total rated power of tractors of less than 20 hp (10⁴ hp)			3	326	1949	2724	3926	4579	5428
Tractor-drawn implements of large and medium type (10⁴)			26	61	131	144	128	113	101
Power-driven rice-seedling transplanters (no.)				11 142	92 578	44 418	17 508	13 605	12 970
Combine harvesters (no.)	284	1789	6704	9164	23 026	34 072	35 861	34 573	30 945
Motor powered threshers (10⁴)			11	103	233	269	323	344	368
Trucks for agricultural use (10³)		4	11	25	97	207	349	426	499

Source: Editorial Board of China Agriculture Yearbook (1984, and for the years 1984, 1985, and 1986. Agricultural Publishing House, Beijing; Chinese Academy of Agricultural Mechanization Sciences (1984). *The General Situation of Agricultural Mechanization in China*. Chinese Academy of Agricultural Mechanization Sciences, Beijing; and Xue Moqiao (ed.) (1985 and 1986). *China Economic Yearbook 1985*, and for 1986. Economic Management Publishing House, Beijing (In Chinese); Editorial Board of China Agriculture Yearbook (1986). *China Agriculture Yearbook 1985 (English Edition)*. Agricultural Publishing House, Beijing; Editorial Board of China Agriculture Yearbook (1987). *China Agriculture Yearbook 1986 (English Edition)*. Agricultural Publishing House, Beijing.

TABLE 5.4   Agricultural mechanization in China, 1952–86

	1952	1957	1965	1973	1979	1982	1984	1985	1986
Area ploughed with tractor-drawn implements (10^4 ha)	14	264	1558	2651	4222	3512	3492	3444	3643
Tractor ploughed area as a percentage of cultivated area		2	15	26	42	38	36	36	38
Area sown with tractor-drawn implements (10^4 ha)				560	1549	1353	1233	1211	1296
Tractor sown area as a percentage of sown area				4	10	9	9	8	9
Area of rice transplanted with transplanters (10^4 ha)				6	24	12	13	13	16
Rice area transplanted with transplanters as a percentage of total rice area				0.2	0.7	0.4	0.4	0.4	0.5
Area harvested with combine harvesters (10^4 ha)				213	384	459	477	475	551
Combine-harvested area as a percentage of sown area				1.4	2.6	3.2	3.3	3.3	3.8

Source: Editorial Board of China Agriculture Yearbook (1984, and for 1985, 1986, and 1987). *China Agriculture Yearbook 1983*, and for the years 1984, 1985, and 1986. Agricultural Publishing House, Beijing; Chinese Academy of Agricultural Mechanization Sciences (1984). *The General Situation of Agricultural Mechanization in China*. Chinese Academy of Agricultural Mechanization Sciences, Beijing; and Xue Moqiao (ed) (1985 and 1986). *China Economic Yearbook 1985*, and for 1986. Economic Management Publishing House, Beijing (In Chinese); Editorial Board of China Agriculture Yearbook (1986). *China Agriculture Yearbook 1985 (English Edition)*. Agricultural Publishing House, Beijing; Editorial Board of China Agriculture Yearbook (1987). *China Agriculture Yearbook 1986 (English Edition)*. Agricultural Publishing House, Beijing.

trucks and cars privately, accounting for one-third of the automobiles used in agriculture. By 1984 the number of trucks owned by the peasants had increased to 186 000, and by 1986 to 318 000 or 64 per cent of the trucks used in agriculture.

The forms of rural ownership that emerged after 1979 were mainly of two types, apart from the state farms which occupied about one per cent of the farm area of the country. These were individual ownership, and collective ownership, which included land managed directly by the collective, and land managed separately under the family contract responsibility system. These new types of land ownership are still developing.

Table 5.3 shows that the increase in the total power rating of the farm machinery of China during the 7 years from 1979 to 1986 was 129×10^6 hp or 71 per cent, and that of the tractors was 44×10^6 hp or 92 per cent. However, where field practices were concerned, only the area harvested with machines showed an increase, and that a small one of 43 per cent. The areas ploughed or sown with tractor-drawn implements, or machines, and the areas transplanted with rice seedling transplanters all decreased over this 7-year period (see Table 5.4). The additional tractor power was, therefore, mainly used in rural transport, and the additional power provided in other types of engines was mainly used in rural industries other than cropping, and for agro-product processing.

These trends were reflected in machinery manufacture: production of machines for field practices decreased, while the production of motors and machines for use in agro-product processing and poultry raising increased rapidly.

In recent years there has been an increased movement of people away from cropping and into other rural industries, and this trend is still in progress. By the end of 1985 18 per cent of the labour force in the rural areas was employed in non-agricultural activities. As a result of this population transfer the cultivated land is gradually being transferred into the hands of capable peasants so that they can form 'specialized-household' family farms (see Chapter 6) where the family will concentrate on producing 'grain', cotton, vegetables, or other farm produce for sale (rather than for home consumption). Most of these specialized household farms are quite small and consist of only one hectare or so of land, though some are a little larger. They differ from the former production teams of the commune type of farming in that the peasants can make decisions themselves, and have the incentive to run their farms effectively. These peasants appreciate the advantages of mechanizing farming practices, but most of their farms are still too small for farm machinery to perform efficiently. Nevertheless, this current system is, perhaps, the better agricultural production system for China for the present. China is still a developing country of a socialist nature, and does not have a strong economic capacity to build a very large farm machinery industry. It is obvious that China cannot have the western type of agricultural mechanization, but must develop her own. However the Chinese government will continue to attach great importance to developing the farm machinery industry, so that both collectivised and individual peasants will be helped to increase production, and also to use more farm machinery when the conditions are right for them to do so.

References

The following texts are in Chinese:

Chinese Academy of Agricultural Mechanization Sciences (1984). *The General Situation of Agricultural Mechanization in China*. Chinese Academy of Agricultural Mechanization Sciences, Beijing.

Editorial Board of China Agriculture Yearbook (1986). *China Agriculture Yearbook 1985*. Agricultural Publishing House, Beijing.

Lu Xianzhou (1963). *The History of Agricultural Machinery Development in Ancient China*. Science Press, Beijing.

Transportation and communications

Zhang Qizong

Transportation

Railways

In 1949 there were only 21 800 km of railway lines open to traffic in China, but by 1985 this total had risen to 52 100 km. In addition, there were also a further 3 760 km of extension lines open which were intended for special use in mining and other industries. Railway transport is now available to every provincial and autonomous region capital except for Lhasa in Tibet. Most of the railway lines, however, run in the east of the country.

The railway has been the most important means of transport so far in China, especially for long distance carriage of large and bulky freight, and also for medium and long distance passenger transport. But in the vast rural areas there are only a few railway lines, and the lines that do run through these areas have few stations. The railways can, therefore, only serve as a partial and indirect means of transport for the incoming requirements and outgoing produce of the country people. In other words, the railway service is limited to long distance transport of bulky consignments, and individual deliveries are made by other means of transport to the various destinations. Nevertheless, the importance of the railways in China is unquestionable because they provide the only economic and practical means of long distance transport of bulky freight, apart from water transport which can only reach a few ports.

Unfortunately, China's railway network today is so sparse and the length so far from adequate to meet the demands, that the service, whether for passengers or for freight, is constantly over-crowded and over-loaded. The railway network is badly in need of new lines, but these require a large investment and in the many mountainous areas of China they require an even larger investment. China's wealth is limited, and the construction of new lines can therefore only proceed on a very small scale; this situation is unlikely to change in the near future. When this is taken into account, it will be seen that the the more than doubling of the length of the rail system between 1949 and 1985 was really a hard-won accomplishment.

Waterways

There were more than 109 100 km of navigable inland waterways in China in 1985, most of which ran through the eastern part of southern China, with the Yangtze River and its tributaries forming the most important water lanes. Western China is known for its aridity, and there is practically no navigable river worth mentioning in that region. North China also has very few navigable waterways, but there are a considerable number in North-East China. As a rule, water transport dominates wherever there is a navigable water course, because it is unparalleled in its economy of operation. This is especially true in regions south of the Yangtze River, where water transport is dominant and a network links all the waterways leading to most of the important cities; the gentle flow of the water facilitates water transport. The prerequisite for water transport is the presence of rivers that are navigable to the destination, or to a place where goods can be relayed a short distance by other means of transport without much trouble. Therefore, inland water transport is not available as part of the major nationwide transport system for, as mentioned above, China's waterways only lead to localized regions in southern China.

Roads

For extensive areas in the rural districts of the country highways unquestionably provide the main means of transport. In the last 30 years there has been a rapid growth in road construction and an ever-increasing length of roads opened to traffic. The roads available for haulage exceeded 960 000 km in 1986. They lead to every corner of the country, and all of the 2137 counties now have their main county town accessible by road, except for Motuo County in Tibet. Roads also link up 90 per cent of the main towns and 70 per cent of the small towns throughout the country. Thus, in spite of half of the total road length consisting of simply built, sub-standard highways, and most of the rest consisting of third and fourth grade roads, which are narrow with many bends, steep slopes, and poor surfaces (270 000 km are earthen, and one-third are impassable on rainy days), the vast expanse of the rural areas is nevertheless accessible by a network of roads. This, too, has been a hard-won achievement.

A serious problem today is the grave shortage of civilian vehicles. There were 2 230 000 trucks in China in 1985; and the number of vehicles increased to only 3 210 000 if all the civilian buses and minibuses were added. Of these only 425 500 were agricultural motor vehicles, a figure far too small to meet the demand. The peasants have been using their farm tractors for hauling all varieties of farm produce and daily requirements on the highway. Most of the tractors they use in this way are walking tractors, and small four-wheeled tractors (Fig. 5.3). Tractors can be used for haulage, but they are designed to suit farm work, and their use as conventional freight carriers on the highway is not always justified. Firstly, there is an unreasonable loss of economy, and secondly, walking tractors are highly susceptible to road accidents on sharp turns and steep slopes. It may, perhaps, be concluded that the use of tractors for carrying freight will only be temporary, and that the situation should steadily improve.

Air transport and coastal transport

Air transport, as yet, has little to do with the rural economy of China, and coastal transport is also of little consequence, apart from the shipment of coal from North China to South China.

Carts, pack animals, and manual transport

On the flat land in North China the transfer of shipments from the highway or a river bank to the rural household, or a grocer's shop in the village, relies heavily on carts pulled by man or animal.

In South China footpaths are spread over less hilly regions and conveyance is by the carrying pole on the shoulder. Pack animals or the

FIG. 5.3 Walking tractor, Wuxi Tractor Factory
Source: Wuxi Tractor Factory. Photograph by Wan Ruixin.

human back are resorted to as a means of delivery in northern and southern China. In the past, pack animals were the chief means of transport on the Yunnan–Guizhou Plateau for both medium and long distance transport.

Freight transported

There has not been an exact figure for the total amount of freight attributable to the country's rural economy, because the necessary survey would be too complex if a complete picture was to be obtained. It is estimated that goods and materials for agricultural production and rural living consigned from urban areas to the countryside amounted to 272×10^6 tonnes in 1983, and that the freight tonnage of rural industry was 400×10^6 tonnes. Another source states that the freight tonnage of 'grain' in 1985 was 107 910 000 tonnes of which 45 030 000 tonnes was transported by train, 37 910 000 tonnes by truck, and 24 970 000 tonnes by water transport. In 1985, the freight tonnage of chemical fertilizers and pesticides was 64 920 000 tonnes; that of manufactured daily articles was 59 570 000; and that of common salt 28 640 000 tonnes. No statistics are available for cotton, various fresh vegetables, fruits, aquatic products, livestock, and poultry. One thing, however, is certain: the volume of freight is continually increasing, and there has been a particularly large increase since the early 1980s. This situation exemplifies the people's growing anxiety for better transport in the countryside — particularly so that the rural economy can be developed further.

Communications

The postal service in China is highly spoken of, even at the international level. This, however, refers to the basic condition in the urban areas. In the countryside there is a different picture. There are comparatively few post offices in the countryside, and even in the places where there are post offices prompt deliveries are prevented by the poor traffic conditions.

According to the statistics, there were 45 017 country post offices in China in 1985, or an average of 20 in each county. In addition there were 135 684 mail boxes scattered through the countryside, of which 14 000 were assigned to agents in the villages, and 186 590 newspaper delivery posts. The long distance delivery of postal matter is generally by air and rail, medium distance delivery by bus or water transport, and short distance delivery by bicycle, or on foot by postmen or postwomen. Postal deliveries between provinces usually take 3–5 days; deliveries over long distances, a week or more. Villages far from a post and telecommunications office are visited by a postman every few days, which delays the delivery even longer. Newspapers and printed matter may arrive in bundles of issues — with no papers arriving between these deliveries.

In 1985 long distance telephone services were available to 94.8 per cent of the townships and towns in the countryside. In the provinces along the coast in the eastern part of China and in the prosperous suburbs of the large cities, it was reported that 20 461 peasants had had private telephones set up at their own expense.

In China, as a rule, the telegraph service is available anywhere there is a telephone service, but a telegram is usually transmitted from the sending post and telecommunications office only to the receiving office in the main county town, or at the district level, where it is decoded and dictated to the destination by telephone. The telegram, now in the form of plain characters in Chinese, is delivered to the receiver by the postman. Thus, a telegram reaches its destination in a couple of days.

References

The following text is in Chinese:
Editorial Board (1987). *China Transportation and Communications Yearbook 1986*. Communication Press, Beijing.

Energy consumption in rural areas

Zhu Zhaoling

The energy consumed in the rural areas comes from two sources: commercial energy supplied by mines or industry, and locally produced or harnessed energy. The former includes electricity, coal, and fuel oil, and is usually called conventional energy. The latter includes biomass such as firewood and crop stalks, hydro-electric power, wind power, and solar energy. Man-power and animal-power are also two common and important forms of energy in China's rural areas, but we will not deal with them here because of their different nature.

Figure 5.4 shows the general picture of energy consumption in China's rural areas. The percentages shown in the diagram are not very accurate, in fact there are other figures on this topic suggested by other scientists. Since China's rural areas are very large and the population is widely dispersed, the gathering, or even the estimation of statistics, is very difficult. Nevertheless, the general picture provided by the diagram is reliable.

Before the end of the 1940s there was little electricity supplied to the rural areas of China, and no modern farm machinery or engine power was used in most of the these areas. Since the beginning of the 1950s there have been major changes. The amount of electricity consumed in domestic use or in production in the rural areas, and the amount of diesel oil consumed in agriculture have increased many-fold in the last 30 years. However, conventional energy still accounts for only a small proportion of the total energy consumed in the rural areas, because of the limitation imposed by the

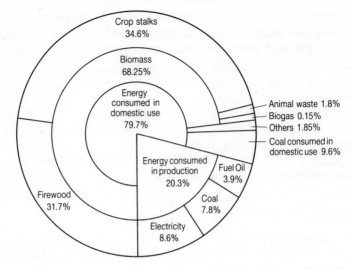

FIG. 5.4   The structure of energy consumption in the rural areas of China.

level of industrialization so far achieved. About 70 per cent of the energy consumed in rural areas still comes from biomass — mainly from firewood and crop stalks — and about 86 per cent of this biomass is used in domestic use. This has an unfavourable impact on the ecological environment of the countryside, which will be discussed below. The conventional energy, which accounts for one-fifth of the total consumption is used mainly as a source of agricultural power.

The total amount of energy consumed in rural areas annually is very large and amounts to some 360 million tonnes of standard coal equivalent, or nearly 40 per cent of the total energy consumption of China. (In order to compare different forms of energy it is common to convert the respective quantities, using different coefficients, into their equivalent of standard coal which has a capacity to produce 7 000 kcal thermal energy per kg). However, the per caput energy consumption for people living in the rural areas is very low and is only about 400 kg of standard coal per year. This partly reflects the fact that some 80 per cent of the total population live in the countryside.

Conventional energy

As mentioned above, conventional energy consists of electricity, coal, and fuel oil, and is supplied mainly by mines and the electricity and oil industries. The electricity and diesel oil supplied to the rural areas is sold at favourable prices, far lower than those for other users. There are also still some small coal pits and small hydro-electric power stations owned by rural people themselves.

The total consumption of coal in rural areas corresponded to 63×10^6 tonnes of standard coal in 1985, which was 17.4 percent of the total energy consumed in rural areas. Of the total coal consumed, 55 per cent was for domestic use, and the remainder was for heating installations in various production processes. It is worth noting that 57 per cent of this coal was provided by the small coal pits which are distributed widely in some parts of the country.

The area covered by the electric power network is not yet very large, and most of the power supplied is used in production processes, including pumping and irrigation. Consequently only a small amount is used for illumination or for domestic electrical appliances. Many peasants' households still depend upon kerosene lamps for lighting. About one-quarter of the electricity used comes from small hydro-electric power stations.

Diesel is the main fuel oil together with some petrol, and these fuels are mainly used in tractors, diesel engines, and trucks. As a result of the development of commodity production in rural areas, rural traffic has been increasing rapidly in recent years and tractors and trucks now form quite a large proportion of this traffic. There has also been an annual increase in the use of modern farm machinery, including pumping and irrigation equipment, which has meant an ever-growing need for fuel oil to operate these items. With the use of mechanized road transport and of mechanized farm machinery and equipment increasing, the demand for diesel oil has been so great that shortages have appeared in rural areas. This demand is nevertheless likely to go on increasing for some time to come.

Firewood and crop stalks

Nearly all of this type of fuel is used directly in the peasants' households, and it is the main form of energy used domestically. Most Chinese rural households obtain their fuel from the crops they grow, or from the trees they grow in small forests, particularly in the hilly areas in South-East China. The heat efficiency of the firewood stoves used widely in the Chinese rural areas is, in general, very low at only 10–15 per cent. This, together with the population growth in the rural areas, has meant that there has been a growing shortage of domestic fuel in the countryside. The demand for firewood has surpassed what can be obtained from reasonable cutting of the firewood forests and deforestation and degradation of the environment have occurred. Measures are now being taken to try to resolve this serious problem — including growing more fuel forests; building more biogas generators in places where it is suitable to do so; and trying to bring about a more widespread use of improved firewood stoves which have a heat efficiency of more than 20 per cent.

Small coal pits and small hydro-electric power stations run by the peasants

The amounts of coal and electricity supplied by mining and the electricity industry for the rural areas are not sufficient to satisfy the demand, and as a result of the new policies introduced with the economic reforms, the Chinese peasants are now enthusiastic to establish their own small hydro-electric power stations and coal pits.

About half of the counties in China have potential hydro-electric power resources which can be exploited to meet part of their requirement for electricity. Most of these resources have a quite small potential capacity, and are not suitable for exploitation via large-scale hydro-electric power stations. However, they can be exploited using small-scale stations. The total installed capacity of small hydro-electric power stations already operating is about 10 per cent of that of all the electric power stations in the country, and the electricity they produce is about one-quarter of that used in the rural areas. Thus these small stations play an important part in rural production.

There are quite large areas, especially in the north and west of China, where there are coal mines. From the end of the 1970s, small coal pits run by the peasants developed rapidly, and increased the national output of coal considerably, while at the same time bringing wealth to the peasants. These small coal pits have advantages in that they can exploit small deposits of coal profitably, which it would be impossible for the larger mining enterprises to do. They also create problems; the major one is that the mining is generally carried out with a lack of scientific design, and safety is not given sufficient attention. In some situations a small pit may destroy other coal pits and create serious accidents.

Other energy resources

Biogas, solar energy, wind energy, and geothermal energy are other energy resources which can be utilized in the rural areas of China. They are, however, only supplementary resources because their utilization is restricted by the natural conditions and they are not always reliable either in time or place.

China has been famous for the development of biogas generators in rural areas in recent years. The peasants construct methane-generating pits, put crop stalks or residues into them, add animal or human excrement, and then wait for biogas to be generated through fermentation. The methane produced can be used for cooking or lighting, and there are some instances where it has been used as a power source in production. The residual slurry is good manure for the crop land.

There are various types of methane-generating pits, designed for different conditions and different purposes. These range from the small ones for a peasant household, to the large ones used as a fuel source on pig farms. However, there is an obvious drawback in that the biogas is produced through fermentation by micro-organisms which have a heat requirement, and do not multiply at low temperatures. This means that the process is more suitable for southern China, where a methane generating pit can produce gas for 6–10 months of the year. But while temperatures are low no methane is generated, and so the process can only be a supplementary energy source. Nevertheless, the process is growing in popularity in the rural areas because of its cheapness. In order to use the methane gas produced from these pits to the best advantage, scientists have developed, and are still developing, various types of gas stoves and gas lamps.

Solar energy is abundant in China. Two-thirds of the country receives solar radiation of more than 140 kcal/cm²/year. Solar cookers, solar heat collectors, and solar driers are the most popular means of utilizing solar energy, and are most suited to places with plenty of solar energy and a lack of firewood, such as North-West China and the Tibetan Plateau.

Wind energy has been utilized in the rural areas for a long time, but using the wind as a power source is more restricted by natural conditions than directly using the sun's energy. Wind power is only used in some areas where the conditions are appropriate for its use in small-scale pumping, irrigation, and power generation — mainly in remote border regions.

In the places where geothermal energy is available, the people are usually able to harness it at low cost. However, such places are few and are not widely or uniformly distributed. Of the sites that are known so far, it is said that about 14 per cent have been exploited for production purposes. The heat is mainly used in green-houses or plastic-film shacks to grow vegetables or to raise fish.

References

The following texts are in Chinese:

Chinese Academy of Agricultural Engineering, Research, and Planning (1981). An Approach to Solve the Problem of China's Rural Energy. Unpublished paper, Beijing.

Chinese Academy of Agricultural Engineering, Research, and Planning (1982). A Study of the Strategy for Rural Energy in China. Unpublished paper, Beijing.

Editorial Board of China Agriculture Yearbook (1986). *China Agriculture Yearbook 1985*. Agricultural Publishing House, Beijing.

Rural Energy and Environmental Protection Office, Ministry of Agriculture, Animal Husbandry, and Fishery (1986). 'Investigations on the status of geothermal energy and wind power in agriculture.' *Journal of Rural Energy*, pp. 20–21.

Lou Xingfu (1985). 'Approach to the estimation and reasonable utilization of firewood forest in China (Part II)'. *Journal of Rural Energy*, p. 19.

The following text is in English:

Editorial Board (1986). *Proceedings of the International Seminar of Rural Energy Technology*. Mechanical Engineering Publisher, Beijing.

6

The rural economy

CHEN DAO, XIE LIFENG, and YAO CHAOHUI

Agriculture and the rural economy as a part of the national economy

In China, agriculture has always been regarded as an important part of the national economy. For the past 30 years agricultural policy has been drafted in accordance with the principle that 'agriculture is the basis of the national economy and "grain" is the basis of that basis'. The background to the formulation of agricultural policy is that China has a population of one billion people, which is more than 20 per cent of the population of the world, but her cultivated area is only seven per cent of that of the world; hence food supply is the main problem to be solved.

The consumption of agricultural products

Agricultural products provide the main necessities of Chinese life. The consumption of agricultural products and of processed goods of agricultural origin per person, is more than 80 per cent of the consumption of all goods, per person, by value. However, the average net increase in the population, even with strict family planning, was 11.7 million people per year during 1981–87. The amount of agricultural produce, especially of 'grain', needed to supply so many people is not only extremely large, but is also steadily increasing. The consumption per person of the main items in recent years is shown in Table 6.1.

From this table it is apparent that it is not a simple matter to meet the demand of a billion people for the main agricultural products. Taking 'grain' as an example, in 1984 the production of 'grain' was a record 367×10^6 tonnes. This was one of the highest levels of 'grain' produced in the world, but the per caput production was still only 357 kg, slightly less than the average level for the world. In the past the supply of 'grain' has been rather tight. For a long period before the economic reforms began in 1979, there had been no free market for the main sorts of 'grain', but a unified standard ration for the city inhabitants, and in the rural areas a ration that differed in standard according to the different production conditions in the different regions.

TABLE 6.1   The average annual consumption of the main agricultural products per person in China in 1978, 1980, and 1985

Item	1978	1980	1985
'Grain' (kg)	195.5	213.8	254.4
Cooking oil (kg)	1.6	2.3	5.1
Pork (kg)	7.7	11.2	14.0
Fresh eggs (kg)	2.0	2.3	5.0
Sugar (kg)	3.4	3.8	5.6
Cloth (m)	8.0	10.0	11.7

Note: The figures include both the goods supplied by the market and the goods produced by the peasants for their own consumption. Cloth includes cotton cloth and cloth made from synthetic fibres.
Source: State Statistical Bureau (1986). *Extracts of the Statistics of China 1986.* Chinese Statistical Publishing House, Beijing. (In Chinese)

'Grain', as defined in Chapter 3, is used in two main ways: directly as food for humans, and as a raw material. The direct consumption of 'grain' as a foodstuff by human beings accounts for 70 per cent of the total production of 'grain'; and this percentage is not likely to change very much in the near future. The use of 'grain' as a raw material includes its use as feed for animals and poultry, and as a raw material for food processing, feed compounding, and for other industries. The demand for these latter uses, especially those for animal husbandry, will increase considerably as animal husbandry is developed. However, for the time being, 'grain' production is more or less sufficient to meet the demand. But if the proportions of meat and eggs in people's diets are to be substantially increased, then the amount of 'grain' used for feed, and in the feed industries, will have to be increased greatly.

If the per caput consumption of 'grain' in China were to increase to 400 kg per year by the year 2000, the total amount required annually would be over 500×10^6 tonnes, at a rough estimate. In 1983 total world exports of cereals were 224×10^6 tonnes, so it would be difficult for the world market to provide China with the additional quantities of cereals of the order required to sustain such a new high level of consumption. Furthermore, China could not afford to pay the foreign exchange necessary for such purchases. Other problems would include insufficient harbour facilities and a lack of adequate transport to handle large cereal imports. Thus, if per caput 'grain' consumption is to be substantially increased, the only way to achieve this seems to be through self-reliance. It is therefore easy to see why the Chinese government has always looked upon 'grain' as one of the most important products in relation to the people's livelihood, and upon agriculture as the basis for the development of the national economy.

The contribution of agriculture to the national economy

In China, in 1985, the proportion of the Gross Output Value of Agriculture (GOVA) in the Total Product of Society (TPS) was 28 per cent — at 458×10^9 yuan in 1624×10^9 yuan. (See Appendix 2 for definitions of these economic terms.) This proportion had dropped considerably from the 1952 level (Table 6.2).

Also in 1985 agriculture had contributed 41 per cent, or 283×10^9 yuan, to the National Income (NI) of 682×10^9 yuan. This percentage fell below 40 per cent in the 1970s but it has risen above 40 per cent again in recent years, although the long-term trend seems to be for it to be dropping (Table 6.3).

In China more than 95 per cent of the funds needed for economic development come from domestic accumulation, and within this agricultural accumulation is an important source. Funds are obtained from agriculture in the following ways: (1) By transferring part of the value of the surplus products produced by the peasants to the state revenue through unequal exchange values between industrial and agricultural products. In China the exchange between industrial and agricultural products is generally unequal in value, though this has been apparent to a different extent in different periods. In the early days of the Communist regime this differential was quite large. In the 1950s the peasants contributed at least 30 per cent of the national revenue through this process of price redistribution, according to calculations based on the price parity between some main industrial and agricultural products in the international market and the domestic market, before the outbreak of the Anti-Japanese War. Since then the government has been continuously adjusting the prices of agricultural and industrial products, but the adjustments have to be confined, to a certain extent, because of the complexity of the price system. In addition, agriculture has to be able to accumulate the capital needed for national industrialization. (2) By the direct taxation of agriculture. In 1985 the direct taxation of

TABLE 6.2   The Gross Output Value of each of the five material production sectors as a percentage of the Total Product of Society (TPS), 1952–85

Year	Agriculture (%)	Industry (%)	Construction (%)	Transport (%)	Commerce (%)
1952	45.4	34.4	5.6	3.5	11.1
1965	30.9	52.0	6.6	3.4	7.1
1978	22.9	59.4	8.3	3.0	6.4
1985	28.1	53.7	10.0	2.7	5.5

Source: State Statistical Bureau (1986). *China Statistics Yearbook 1986*. Chinese Statistical Publishing House, Beijing. (In Chinese)

TABLE 6.3   The Net Output Value of each of the five material production sectors as a percentage of the National Income (NI), 1952–85

Year	Agriculture (%)	Industry (%)	Construction (%)	Transport (%)	Commerce (%)
1952	57.7	19.5	3.6	4.3	14.9
1965	46.2	36.4	3.8	4.2	9.4
1978	35.4	46.8	4.1	3.9	9.8
1985	41.4	41.5	5.5	3.5	8.1

Source: State Statistical Bureau (1986). *China Statistics Yearbook 1986.* Chinese Statistical Publishing House, Beijing. (In Chinese)

agriculture reached 4.205×10^9 yuan, equivalent to 2.3 per cent of the state revenue for that year. The government has consistently adopted a policy of taxing agriculture lightly, and hence the proportion of the taxation on agriculture in the revenue has been dropping — from 14.7 per cent in 1952 to 2.3 per cent in 1985. In a comparison between 1984 and 1985, agriculture accounted for 29.1 per cent of the growth in NI, but for only 2 per cent of the increase in the state revenue. That is, the state revenue claimed very little of the portion of the growth in NI contributed by agriculture, which speaks volumes for the government's endeavour to support the development of agriculture. (3) By taxing the processing, transportation, and marketing of agricultural produce. This is indirect taxation of agriculture and the sums raised are very large, but it is difficult to make a precise calculation. There should be no doubt, however, that the ups and downs of agriculture directly affect this part of the state revenue.

Over the longer term, the proportions of the GOVA and the Net Output Value of Agriculture (NOVA) in the national economy have been gradually declining, which is a trend consistent with that of many other countries. For the time being, however, these proportions are still large and agriculture is still playing a role of fundamental importance in the national economy. In recent years, the steady growth in agriculture has brought prosperity to the rural economy as well as to the nation as a whole. An obvious aspect of this is that the peasants have been persistently setting up non-agricultural enterprises encompassing rural industry, construction, transport, commerce, etc., and these have shown rapid progress. In 1980 the Rural Total Product of Society (RTPS) accounted for 32.7 per cent of the Total Product of Society (TPS). This percentage had risen to 38.9 per cent in 1985, and it can be predicted that it will rise further. However, the rise can not be very fast because centralized large-scale industries are also being built and developed, so the urban industries are still dominant in the national economy.

The relationship between agriculture and industry

China's agriculture is closely related to her industry. The raw materials for industry, except those produced by industry itself, are largely provided by agriculture. Agriculture supplies the food processing industry, textile industry, breweries, tanneries, pharmaceutical industry, and other industries, with 'grain', cotton, vegetable oil, sugar, cocoons, tea leaves, wool, hides, medicinal herbs, and many other products. In 1985, that part of light industry which used agricultural products as raw materials produced 33 per cent of the total value of all industrial production, while that part which used non-agricultural products as raw materials produced only 17 per cent.

In 1952 the value of the production from light industry using agricultural products as raw materials was 88 per cent of the total value from all light industry (see Table 6.4). After the 1960s the development of the petrochemical industry made available chemical fibres, plastics, synthetic materials, and other newly invented materials which could be used in light industry. The use of these oil-based materials became increasingly more widespread, and induced a revolutionary change in the composition of the supply of raw materials for industry. By 1985 the value of the production from light industry using agricultural products as raw materials had fallen to 66 per cent of the total from all light industry.

Despite the growth of the petrochemical industry, the market provided by industry for agriculture did not vary very much during the period 1949–78, due to an insufficient supply of agricultural products to the market. However, in recent years the functioning of the market mechanism operating since 1979 has brought changes. In 1984 there was a record-breaking harvest

TABLE 6.4 The use of agricultural products as raw materials in light industry in 1952, 1978, and 1985

Year	The value of production from light industry using agricultural products as raw materials (10^9 yuan)	The first column as a percentage of the total value of production from all light industry (%)
1952	19.4	88
1978	123.6	68
1985	295.7	66

Source: Department of Rural Statistics, State Statistical Bureau (1984). *The Glorious Achievements of Chinese Agriculture*. Chinese Statistical Publishing House, Beijing; State Statistical Bureau (1986). *China Statistics Yearbook 1986*. Chinese Statistical Publishing House, Beijing. (In Chinese)

and the market mechanism at that time was imperfect, which gave people a false impression that agricultural products were in surplus. Then a change was made in agricultural policy, which in turn caused a slight fall in 'grain' production, and a drastic decrease in cotton production. However, since 1986 the peasants have been encouraged to grow more 'grain' and cotton again. These dramatic changes have indicated clearly that agricultural products are in fact still in short supply. One of the reasons for this continuing shortage of supply in the market is that the demand for agricultural products by industry is steadily increasing. The food industry showed an annual average increase in demand of 9.4 per cent during 1979–85, compared with the annual average increase of 6.3 per cent in 1953–78, and this is reflected in the market for agricultural products.

Agriculture supplies industry with a large quantity of raw materials; in turn, agriculture and the people of the rural areas provide an extensive market for the products of industry. There are some 191 million peasant households in the rural areas, and these are now integrated into many other parts of the national economy (see Table 6.11) — an integration that is clearly reflected in the increased importance of cash income and expenditure within the overall income and expenditure of the peasant households (see the last section of this chapter). Some decades ago the subsistence peasants did not rely much on goods purchased from the market because they were short of money and could not afford to buy such goods, but by 1987 the peasants were spending some 65 per cent of their living expenditure on market purchases (see Table 6.24), and the value of retail sales in the rural areas (of consumer goods and goods for agricultural production) was fairly stable at around 59 per cent of the value of all retail sales in China by the mid-1980s (Table 6.5).

Expenditure on items for use in agricultural production has mainly been on farm machinery, chemical fertilizers, new hybrid seeds, plastic film, and similar goods. However, since 1979 the rural population's expenditure on consumer goods has increased at an even faster rate than its expenditure on items for agricultural production, and expenditure on durable consumer goods has become an increasingly important feature of the expenditure on consumer goods (see Table 6.21). That expenditure on consumer goods should increase at a faster rate than expenditure on goods for agricultural production is to be expected, because of the movement of people out of agriculture and into non-agricultural activities. The figures in Table 6.5 do not include the additional goods purchased for use in the township enterprises. However, one thing is incontrovertible and that is that the rural areas are now providing an extensive market for industry.

At present about 60–70 per cent of the products produced by light industry are sold in the countryside. Products produced by heavy industry, such as motor vehicles, tractors, large and medium sized farm machines, and farm implements, also find a market in various rural areas. A steady increase in

TABLE 6.5   The retail sales of commodities in China, 1952–85

Year	Retail sales in China					Retail sales in rural areas as a percentage of all retail sales in China
	Total	Consumer goods		Goods for use in agricultural production		
	(10^9 yuan)	(10^9 yuan)	(%)	(10^9 yuan)	(%)	(%)
1952	27.68	26.27	94.9	1.41	5.1	54.6
1965	67.03	59.01	88.0	8.02	12.0	49.4
1978	155.86	126.49	81.2	29.37	18.8	52.0
1979	180.00	147.60	82.0	32.40	18.0	54.7
1980	214.00	179.40	83.8	34.60	16.2	55.6
1981	235.00	200.25	85.2	34.75	14.8	56.3
1982	257.00	218.15	84.9	38.85	15.1	57.6
1983	284.94	242.61	85.1	42.33	14.9	58.6
1984	337.64	289.92	85.9	47.72	14.1	59.2
1985	430.50	380.14	88.3	50.36	11.7	58.5

Note: The 'Consumer goods' in this table are food, household goods, and fuel.
Source: State Statistical Bureau (1986). *China Statistics Yearbook 1986*. Chinese Statistical Publishing House, Beijing. (In Chinese)

the peasants' incomes is a prerequisite to maintaining these extensive markets, and the Chinese government is aware of this situation and pays much attention to it whenever the peasants' incomes fluctuate.

The import and export of agricultural products

The export of agricultural products makes up a substantial proportion of China's export trade. The items exported include 'grain' (particularly rice and soya beans), peanuts, fresh eggs, live pigs, poultry, fruit, aquatic products, cotton yarn, flue-cured tobacco, and a wide range of other plant and animal products and by-products, and processed items. Among the more traditional items exported are goat skins, bristles, rabbit hair, resin, honey, tung oil, green tea, Chinese herbal medicine, freshwater pearls, and hand-drawn work; the exports of some of these more traditional items can comprise 60–80 per cent of their world trade.

In 1985, China's foreign trade organizations exported unprocessed and processed agricultural products to the value of US$11.5 × 10^9, 42 per cent of the value of total exports for that year. Although this was a substantial contribution to exports, the general trend over the last three decades has been for industrial and mineral exports to form an ever increasing proportion of total exports (Table 6.6). Due to the more rapid real increase in non-agricul-

TABLE 6.6   The average annual values of total Chinese exports, and of the exports of unprocessed and processed agricultural products, 1950–86

| Period | Total exports | Exports of agricultural products | | | |
| | | Un-processed | | Processed | |
	$(10^9$ US\$)	$(10^9$ US\$)	(%)	$(10^9$ US\$)	(%)
1950–59	1.320	0.589	44.6	0.422	32.0
1960–69	1.939	0.620	32.0	0.774	40.0
1970–79	6.622	1.951	29.5	2.253	34.0
1980–86	22.899	3.961	17.3	6.329	27.6

Source: State Statistical Bureau (1989). *Selected Superior Papers on Statistical Analysis*. Chinese Statistical Publishing House, Beijing. (In Chinese)

tural exports, the percentage contribution to the export trade of the agricultural exports, particularly that of the unprocessed agricultural products, has been falling over the long term. However, in volume terms the exports of various individual agricultural products increased slightly during the first half of the 1980s. In 1980 and 1985 the dollar values of the unprocessed agricultural products exported were US\$3.42 and 4.53×10^6, respectively, and of the processed agricultural products US\$5.60 and 6.97×10^6.

Unprocessed and processed agricultural products form only a small part of total Chinese imports. In 1985 the value of the agricultural products imported was US\$$2.96 \times 10^9$, or 7 per cent of total imports. The agricultural and forestry products imported to a value in excess of US\$$100 \times 10^6$ in the same year were cereals, logs, wool, sugar, and natural rubber — in descending order of value.

The import and export of 'grain' has long been a trade of considerable magnitude. During the late 1970s and early 1980s, the yearly import of 'grain' into China was in excess of 10×10^6 tonnes, and in international terms was third after the imports of the USSR and Japan in value. This contrasts with the period before the 1960s when China was a small-scale exporter of 'grain' (Table 6.7).

As explained above, China's requirement for 'grain' is so large that it would be unrealistic for the country to rely on the international market for a very major portion of its supply over the long term. However, the need to adjust the availability of the different types of 'grain', the surpluses and deficiencies in the supply, and the regulation of the internal market, means that it has become quite normal to import 'grain' at one time and to export it at another.

TABLE 6.7 The 'grain' imports and exports of China, 1952–85

	1952 (10⁴ tonnes)	1957 (10⁴ tonnes)	1965 (10⁴ tonnes)	1979 (10⁴ tonnes)	1980 (10⁴ tonnes)	1984 (10⁴ tonnes)	1985 (10⁴ tonnes)
Exports	153	209	242	165	162	319	933
Imports		17	640	1236	1343	1041	597
Imports of wheat		5	607	871	1097	987	538

Note: 'Imports of wheat' are included in 'Imports'.
Source: State Statistical Bureau (1986). *China Statistics Yearbook 1986*. Chinese Statistical Publishing House, Beijing. (In Chinese)

The agricultural labour force

In 1985 the labour force of China was recorded at just under 500 million people. Of this total, 312 million were engaged in farming, forestry, animal husbandry, fishery, and water conservancy — in the national statistics relating to labour this group of activities has been called the primary industries. In recent decades the proportion of the labour force working in the primary industries has been decreasing as the country has become more industrialized: in 1952, 84 per cent of the labour force worked in the primary industries, by 1985 this figure had fallen to 63 per cent (Table 6.8).

Although the proportion of the total labour force engaged in the primary industries is falling, the actual numbers of people so engaged are still increasing, due solely to the natural increase in the population (see Table 6.8). However, the increases that are occurring in the numbers of people engaged in the secondary and tertiary industries are coming both from the natural increase in the population and from a transfer of people from the

TABLE 6.8 The distribution of the labour force between the different economic sectors in China, 1952–85

Sector	1952 (10⁶)	1952 (%)	1965 (10⁶)	1965 (%)	1978 (10⁶)	1978 (%)	1985 (10⁶)	1985 (%)
Primary	173	83.5	234	81.6	294	73.8	312	62.5
Secondary	15	7.4	24	8.4	61	15.2	105	21.2
Tertiary	19	9.1	29	10.0	44	11.0	82	16.4

Source: State Statistical Bureau (1986). *China: A Statistical Survey in 1986*. New World Press, and China Statistical Information and Consultancy Service Centre, Beijing. (In English)

labour force of the primary industries into the labour forces of these other industries.

After the beginning of the economic reforms in 1979 and the improvements in the incentives for the peasants, the problem of surplus labour in the rural areas began to emerge. Restrictions on the activities of the labour force were abolished, and the government permitted and even encouraged the peasants to work in non-agricultural sectors in the rural areas. One form of encouragement is the recent policy of 'leave the land but stay in the village', under which new small towns are set up and the development of new enterprises is encouraged, so that some of the surplus labour can become involved in non-agricultural occupations while remaining within the rural areas. As a result a part of the agricultural labour force has shifted into other sectors, but this change has proceeded unevenly in both speed and extent in different parts of the country. In the rural areas of the coastal regions in eastern China, 27 per cent of the labour force is engaged in secondary and tertiary industries, while in the rural areas of the west of China the figure is only 10 per cent.

Given that the area of cultivated land cannot be expanded, a continuing increase in the labour force engaged in the primary industries will certainly hamper improvements in agricultural productivity per person. For example, the per caput production of 'grain' in the mid-1970s showed no increase over that of the late 1950s. When no more job opportunities for surplus rural labour can be found in the cities and non-agricultural sectors, agriculture accepts a substantial number of the people involved at the expense of the efficiency of production. According to the rural basic statistics, in 1985 some 67 million people living in rural areas were engaged all the year round in non-agricultural activities, and this represented about 18 per cent of the total labour force of the rural areas (see Table 6.11).

Rural policies from 1949 to the present

Land reform and the agricultural cooperative movement

For a long period before the founding of the People's Republic of China, the rural economy of China had been dominated by a feudal system of land ownership. Some 70–80 per cent of the land was owned by only 10 per cent of the agricultural population, that is by the landlords and rich peasants. The poor peasants rented most of their land from the landlords.

As in many other countries, land reform in China was brought about as the result of political activity. During the 10 years 1927–36 the Communist Party of China carried through a revolution in land ownership within its Jiangxi Soviet, as a part of the wider revolutionary activities in which it was then involved. The Party deprived the landlords and rich peasants of their property and distributed it to the poor peasants and farm workers who had little or no land. During the Anti-Japanese War a policy of reducing rent and interest was followed, to encourage a unified approach to fighting the Japanese invaders. After the surrender of the Japanese the peasants eagerly demanded land, so the Communist Party again adopted a policy of confiscating the land of the landlords and rich peasants and distributing it to the poor peasants and farm workers.

Then, in 1950, in place of the policy of reducing rents and interest, the government published the Land Reform Law of the People's Republic of China, and in the winter of that year land reform was carried out on an unprecedented scale all over the country. As a result of this movement the peasants with little or no land, who were 60–70 per cent of the agricultural population, were given the ownership of some 700×10^6 mu (47×10^6 ha) of cultivated land, together with houses, farm implements, and farm animals. By the end of 1952 nationwide land reform was more or less complete.

The cooperative movement began shortly after the completion of the land reform. In December 1951 the Central Committee of the Communist Party drafted the Resolution on Mutual Help and Cooperative Transformation for Agricultural Production and implemented it on a trial basis. The background to the Committee making this resolution was as follows. When the peasants had obtained their own land, own draught animals, and other means of production, and their households had become economic units, their initiative increased greatly. Many of the peasants were so satisfied they did not want to see any further changes in their small-scale household farming. However, some of the poor peasants were still economically disadvantaged and short of labour and the means of production. Hence the hiring of labourers, usury, lending, mortgaging, pawning, and the buying and selling of land came into being in some places; while in other places various cooperative activities appeared, such as the exchange of labour, mutual help, and organizing mutual-aid teams. The above resolution was made in order to lead the peasants into mutual help and cooperation.

To begin with, the general approach was to form either temporary mutual-aid teams, or year-round mutual-aid teams. The temporary teams usually consisted of 3–10 households, and they mainly provided mutual help in various aspects of agricultural production; the year-round teams had a more fixed membership and usually drew up a production programme, a system of work points, and a scheme for the arrangement or assignment of the work. Some of the year-round teams bought farm implements and draught animals from time to time in small numbers, as their communal property, but in general the means of production in such mutual-aid teams, such as land and farm animals, were privately owned, and the recruitment of members was entirely based on the principle of voluntary participation and mutual benefit.

At the same time, the Central Committee of the Communist Party began, in the spring of 1950, to try out experimental, primary agricultural producers' cooperatives in which the peasants' land was accepted as the investment, while the farm animals and large farm implements remained privately owned but under common usage. These cooperatives had to pay bonuses and interest to the owners for the use of their land and means of production.

The upsurge of the primary agricultural producers' cooperative movement was in 1955. In June of that year the number of cooperatives reached 630 000 with a total participation of 16.9×10^6 households. In the following year, 1956, advanced agricultural producers' cooperatives came into being, and the private ownership of the land was converted into collective ownership by the cooperatives; thus payment of interest on the investment of land in the cooperative ceased. The other means of production were also transferred into collective ownership after the values of the items were established and payments made to the owners. Members of the cooperatives took part in the work of their cooperative collectively on the basis of the work points system, which embodied the principle of distribution according to work done. Wages were paid according to the respective work points of each individual. Household production was negated, but a plot was given to each family for its private use, and small-scale sideline production was also permitted. Thus the cooperative transformation of agriculture was implemented in a winter and a spring instead of over 15 years, as originally planned.

Two years later, under the principle of 'large organization and public ownership', the advanced agricultural producers' cooperatives were united into people's communes which functioned as economic as well as administrative units. This was followed by the establishment of a three-level administrative system, consisting of the commune, the production brigade, and the production team. The people's communes and the three-level administrative system lasted 20 years.

While the movement toward centralized control of land and labour was in progress, the government exercised unified purchase, and fixed the price for the state purchase of the main agricultural products in order to consolidate the leading position of the state-run enterprises over the market. In 1953, 'grain' and edible oil had to be sold, at prices fixed by the state, to specialized departments through which these agricultural products were then sold to the consumers. In 1954, the same policy was applied to cotton and cotton cloth, and a year later the prices of live pigs and of important raw materials for industry were unified. The state also fixed the quotas of produce the peasants had to sell to the state each year. These were called the purchase quotas. To facilitate the implementation of this policy the government also narrowed the difference in prices between different localities and abolished seasonal differences in prices. After this, the government was gradually able to monopolize the purchase and marketing of agricultural products, and the country fairs slowly began to wither.

The people's communes

The period of the people's communes was from 1958 to 1979. At the beginning of that period the production brigade, the second level in the three-level system, was a large economic body equivalent to the original advanced agricultural producers' cooperative, and consisted of many production teams. All the land, farm animals, farm implements, and other means of production were under the ownership of the brigade. The proportions of the products and income handed over to the production brigade by each production team were in accordance with a plan, and the products and income produced by brigade-run units belonged to the brigade. It had the full right to dispose of the products and income, and it assumed the sole responsibility for profits and losses. Then in 1962 the production team became the accounting unit. It was small in size and easy to manage, therefore it became the basic economic unit in the commune system.

Within the commune, land was public property but under the control of the production team. Small plots were shared out to the households according to the number of persons in the family for their private usage, such as growing vegetables for their own food. The immediate ownership of the plots, however, still rested with the team.

When a person went to work his or her job was assigned by the team leader and work points were recorded for each member according to his or her daily labour. The grades of work points were related to the different elements within the labour force, that is, full male labour, full female labour, half male labour, half female labour; and the work point was the measure for the distribution of income. During the busy farming seasons of spring and autumn, when the farm work was intensive, special measures were sometimes taken, such as contracted stages of work, so that a whole task might be completed at the right time. During the slack seasons the teams and brigades concentrated their labour on irrigation and drainage projects, embanking lands from rivers and lakes, reclaiming waste land, and other works of a capital nature. For large-scale projects, such as building highways or reservoirs, the commune would gather labour from the production teams to do the job. The main task of the team, though, was cropping, especially the growing of field crops such as rice, wheat, and maize. In both northern and southern China it was essential to cultivate these crops to provide 'grain' for both self-consumption and delivery to the government.

The first call on the agricultural produce was to satisfy the state levy and the state purchase quota. These were fixed by the government according to the area of land controlled, and were changed little from year to year. No payments were received in return for the state levy, so it was equivalent to an agricultural tax. State purchase was the purchase of the peasants' 'grain' by the government at prices fixed by the government. Quotas for state purchase had to be filled but in lean years the teams most affected by low yields would be exempted from a part of the state levy and state purchase. If the exemption was not sufficient to alleviate the situation, the government would transfer 'grain' from other places to maintain the peasants' livelihood. Marketable 'grain' was sold to the government departments which dealt with 'grain', through the unified state purchase system, and the marketing of 'grain' at country fairs was prohibited. The planted areas of the main cash crops such as cotton and oil crops were quite large, but the produce still had to be sold to the government. Throughout this commune period, from 1958 to 1979, the prices at which agricultural products were bought and sold were fixed low and generally remained unchanged; this was particularly so in the years 1966–78. During this latter period the country fairs were also gradually abolished through a series of local proclamations.

The method of distributing income at team level was as follows. The production team first calculated the total yearly income which was derived mainly from the value of the 'grain' crops and cash crops. Next the production expenses (excluding payment for the labour force) were deducted from the total income, and then the net income was distributed to each individual according to his or her total work points for the year. Items for subsistence, such as 'grain' and cotton, paid by the team to individual peasants during the year were evaluated in terms of money and deducted from the yearly income.

Replacement of the people's commune system began in 1979 and steadily progressed; now collective production systems are to be found in only a few places in some economically developed regions. Instead, the household contracted responsibility system, linking income with output, has been widely adopted throughout the country in recent years, and at present some 95 per cent of rural households are carrying on agricultural production under this system.

The household contracted responsibility system, linking income with output

At the time that the agricultural producers' cooperatives and people's communes were active, different types of agricultural responsibility systems emerged in practice. These included the contracting of a stage of work, contracting an amount of produce, and working separately by groups on contracted work with income linked to output. However, adoption of the household contracted responsibility system, linking income with output, only became widespread in the early 1980s. Of the different versions of the household contracted responsibility system, the one that the peasants like most is the 'quota contracted to household', which can be managed by the peasants themselves.

According to an investigation in 1985 of 10 481 former production teams (typical of rural areas), 9987 had changed to quota contracted to household, accounting for 95.3 per cent of the total; 402 had changed to production contracted to household, accounting for 3.8 per cent; 25 had changed to quota contracted to group, accounting for 0.2 per cent; 45 had changed to production contracted to group, accounting for 0.4 per cent; four had changed to a stage of work contracted, accounting for 0.04 per cent; and 18 others accounted for 0.2 per cent (*China Agriculture Yearbook 1986*).

Quota contracted to household is a household contracted responsibility system which links income to output. In order to introduce this type of responsibility system, the right of use of the collective land had first to be transferred to the individual peasants. That is, the land of the production team had to be distributed to each member of the team. The team included everybody — old men, old women, and infants in addition to the working-age men and women — and each received his or her share. In general, this distribution was carried out evenly to each member, but sometimes the able bodied adults who did the actual work were given a larger share. Because there are differences in the quality of the land, and the peasants regard the land as very important, the distribution of the land to every household had to be handled as carefully and reasonably as possible, and this has resulted in a scatter of the land contracted to each household. The investigation in the rural areas in 1985 (see above) indicated that the average amount of land per household was 8.35 mu (0.56 ha) and this was scattered in 9.7 plots. The peasants were quite satisfied and regarded this as fair, in spite of the inconvenience it caused.

The ownership of the land remained with the production team and was collective in nature. The government stipulated that it was not permissible to buy, sell, rent, or use the land for any other purpose. The production team should make, within a fixed number of years, any necessary adjustments to the distribution of the land that become necessary following changes in the number of family members or family labour force. To encourage the peasants to make long-term investments in the land, the government suggested prolonging the fixed number of years of the contract — that is, the right of use of the land should endure for not less than 15 years.

The draught animals and large farm implements originally owned by the production team were shared out to groups of households or individual households after the monetary value had been established. Any outsider wishing to use any of these items could hire them in exchange for human or animal labour or for payment in money. At present there are instances where contracted households are uniting on a voluntary basis to form partnerships to buy means of production, such as a draught animal, a tractor, or product processing machinery.

With these changes have come changes in management responsibilities. All matters relating to the management of production and marketing, formerly the responsibility of the production team, are now taken care of by the household. The production team was only left with responsibility for making land contracts, for making yearly purchase contracts with the peasants for agricultural products, for taking care of the operation and maintenance of the medium and large scale irrigation schemes, and for improving the conditions of the peasants. The production teams themselves have now almost entirely been replaced by the township economic committees (see next section). Contracted households work separately and on a small scale. All they have to do is hand over to the government a set amount of agricultural produce according to their contract; and after that they can make their own decisions about what to plant and how much to plant. In this way the peasants' labour and interests are brought together and they are responding with enthusiasm and new found ability.

There are still some groups of people who follow collective production systems, though the units involved are no longer called people's communes or production brigades. Most of these units are to be found in city, suburban, or other economically advanced areas, and only a few per cent of the people in the rural population are involved in these collective production systems.

The key point of the quota contracted to household system, is that of 'ensuring the portion for the state, turning over the portion to the collective, and the portion left, whether large or small, belongs to the peasants themselves'. First of all the peasant household has to fulfil its duties of turning over and selling agricultural produce to the government as stipulated in the contract, and of paying agricultural tax. Then it has to provide to the collective a certain amount of cash or materials as its contribution to the accumulation fund, public welfare fund, and fund reserved for other purposes. The remainder of the household's produce and income is its own. The members of the household can sell their surplus 'grain' and other agricultural products to the state, or on the free market, entirely at their own choice. Prices of agricultural products have been changed since the government transferred the right of use of the land and management of production to individual peasants. In 1979 the government raised the purchase prices of 18 items, including 'grain', cotton, and oil seeds, by 22.1 per cent, and after that raised the prices of cotton, oil seeds, and other items again for purchases above quota. These measures stimulated the peasants to further efforts. Meanwhile the peasants were once again permitted to sell their agricultural products and by-products at country fairs.

The problem with the quota contracted to household system is that the land of a household is scattered in many small plots, and this is not convenient for production. In addition, the per caput area of cultivated land is very small, and each household undertakes its cropping activities independently. All this means that labour productivity is very low. The government is now promoting the transfer of land from those peasants who wish to have it transferred, and from those who may be less skilled, into the hands of the skilled peasants, and is now permitting peasant households to contract out their land for payment, to return their land to the production team, or to transfer their land to other households.

New changes in the rural economy

Economic organization

The most basic management unit in the Chinese rural areas is the contracted household now that the contracted household responsibility system has been adopted. The people's commune system has been dismantled and replaced by township economic committees. The latter are in charge of land use management, and the making and implementing of contracts, while production, marketing, and distribution are left in the care of the peasants.

It is quite clear that the household is a unit which is too small and too dispersed. However, questions about what form of economic cooperative, what kind of rural service system, and what kind of economic committee, should be adopted, and the appropriate sizes for these bodies from a management point of view, are still being studied and discussed by the authorities concerned, with the aim of encouraging the growth of production in a manner that is in keeping with the wishes of the peasants and which does not stifle their initiative.

At the People's Congress in April 1987, the Regulation of the Organization of the Villagers' Commissions was discussed. It is no secret that weaknesses exist in the work at the basic level in the rural areas, and villagers' commissions were intended to be organized at the village level to improve the situation. The preliminary assumption was that the villagers' commissions would be self-governing organizations of the local people. The villagers would be organized so that they could perform better, as masters and in managing financial affairs, while at the same time still being able to use their initiatives. The villagers' commissions are intended to be political as well as economic organizations, but their functions, and the procedures for setting them up, are still under discussion.

Rural economic unions

In places where the household contracted responsibility system was put into practice at an early stage, rural economic unions also came into being early. The characteristic of these unions is joint management. This joint management can be between state-run economic organizations and collectives or households; between collectives and collectives; between collectives and households; or between households and households. According to the 1985 investigation, 3.2 per cent of the total households had joined rural economic unions, involving 3.4 per cent of the rural labour force (*China Agriculture Yearbook 1986*). Of the various forms the unions could take, those between households and households made up the majority, with many of them being between relatives and friends.

There are different kinds of unions engaged in nearly all types of business; they are not confined to a definite region or to a definite trade. However, unions engaged in the agricultural products processing industries and in transport are the most common, with 20.3 per cent of the total being engaged in the former, and 13.5 per cent in the latter (*China Agricultural Yearbook 1986*).

When a union is formed, individuals buy shares; shareholders are paid dividends according to the number of their shares. About half of the capital of the unions is self-raised, and the labourers receive dividends on their shares in addition to their wages. In general, though, these unions are mostly loose in structure and unstable.

Specialized households

Specialized households are new economic entities which have appeared only after the adoption of the household contracted responsibility system. The policy of the government is now to encourage some of the peasants to get rich first, rather than to aim at all proceeding at an equal pace. These peasants might make use of their better resources, and/or their technical expertise, to undertake certain kinds of specialized production which yield good rewards. Under this policy, there have emerged, and are still emerging, numbers of specialized and 'stressed' households in the rural areas. The 'stressed' households were households treated as more important by the government, but this was a previous idea and now the concept of the specialized households is the more important of the two.

There are no universal criteria for defining specialized and 'stressed' households, but judging from what has happened in practice there are certain common elements which can be found in specialized or 'stressed' households in different localities. Firstly, of course, they are all economic entities based on household management. Secondly, compared with common households, more emphasis is placed on a high commodity rate for the main product; thirdly, a higher proportion of the labour force and of labour time is devoted to producing the main product; and fourthly, a higher proportion of the total income is derived from the main product. (The commodity rate is the percentage of the product that is sold on the market, rather than consumed by the household itself). In most cases these last three criteria of commodity rate, specialized labour rate, and specialized income rate, are within the range of 30–70 per cent. Those households which failed to reach these criteria but which were actually engaged in specialized production were called 'stressed' households, but due to its vagueness the term 'stressed' household has not been used since 1984.

There are two kinds of specialized households, the contracted specialized households and the self-managing specialized households. The contracted specialized households are those which have contracted land, fish ponds, or township enterprises from collectives. These peasants are able men or women, or they have particular experience in technical work. Because the resources they are using under contract are the properties of the collectives, they can only have a share of the income, as stipulated in the contract. Since most of these specialized households engage in activities such as cultivating field crops and raising silkworms, they are at risk from bad weather conditions.

The self-managed specialized households are in a better situation because most of them are engaged in transport, poultry raising, or agricultural products processing, where the risks are much less. These households have already provided many jobs for the surplus rural labour force: it is said for up to 20-25 per cent of this surplus. All of the specialized households cultivate small plots of land, as other peasants do, however, the income they derive from this kind of activity is only supplementary to their main income.

The development of specialized households is different in different regions. It is most noticeable in places where the collective economy has not been so powerful in recent years and where an active commodity economy has appeared.

Rural economic services

There are a number of service units in the present-day Chinese rural areas. Some of them are state-run, such as the farm machinery service stations, the seeds corporations, and the veterinary centres. The remainder, such as the trade associations and the specialized service households, have been established by the peasants themselves, and the peasants are also trying to restore peasant-run basic supply and marketing cooperatives, and credit cooperatives.

A trade association is a unit that is something like an economic union which also has the characteristics of a technical service. For example, a cow association is responsible for offering technical guidance and knowledge, disease prevention services, and milk processing and marketing services to specialized dairying households. The extent to which trade associations have developed is related to the characteristics of the products produced in a region. Generally, in places where large amounts of staple products are produced the trade associations are large and widely dispersed, examples are the tea associations in Zhejiang province and the orange associations in Sichuan province.

The basic supply and marketing cooperatives were economic organizations originally set up by the members themselves with pooled funds, but they later became state owned. In 1982 it was decided that cooperatives should again be run by the peasants, and after this decision the peasants joined the cooperatives as shareholders: by 1985 some 85 per cent of rural households had joined the cooperatives. Besides their supply and marketing activities, these cooperatives also provide services relating to information, capital, materials, technology, processing, storage, and transport for their members.

The rural credit cooperatives, which provide peasants with credit services, were set up by local peasants with pooled funds in the 1950s. At present they are under the supervision of the China Agriculture Bank and form its basic branch offices, but they are not state owned. The sources of the capital of the rural credit cooperatives include the money paid for the shares by the members, the deposits received, and the loans provided by state banks. The money is used to help the collective economic units, the township enterprises, and the individual peasants to solve financial problems of production or living. The profits made by the credit cooperatives are partly allocated as dividends to the shareholders, and partly put into an accumulation fund and used to expand the funds available for loan.

More peasants have joined the credit cooperatives since 1979 and they are now quite popular in the rural areas. At present about 80 per cent of all rural households have some relationship with a credit cooperative. At the end of 1985 the total deposit balance of all the credit cooperatives was 72.40×10^9 yuan, and 78 per cent of this had been deposited by the peasants. The total for the loans outstanding at that time was 61.81×10^9 yuan, and 90 per cent of this had been lent to rural enterprises and rural households (*China Agriculture Yearbook 1986*). Thus, the money comes from the people and is used by the people.

The difference between the state banks and the credit cooperatives lies in the fact that the latter mostly receive current deposits and make short term rural loans. In the second half of the 1980s the credit cooperatives have provided considerable financial aid towards the development of the poor or backward regions.

Rural sideline activities and township enterprises

Rural sideline activities

Agriculture in its broad sense comprises crop cultivation, animal husbandry, forestry, fishery, and sideline activities. The first four sectors have already been discussed in Chapter 3, but sideline activities are a little different in nature and will be dealt with here. They may be grouped into three categories: gathering wild plants, hunting wild animals, and handicrafts. The gathering of wild plants includes gathering wild oil-bearing plants, wild fibre plants, and wild starch-bearing plants, which can be used as raw materials for the chemical industry or as medicinal herbs. Because China is a mountainous country, rich in resources of wild plants, the peasants can profit from gathering these at their leisure. In some places in the mountains there are also wild animals which can be hunted; but the hunting must not be too intensive or the resource will be harmed. Handicrafts include weaving, carpentry, masonry, transport, repairing, and other small scale activities. Those engaged in sideline activities include peasants with other main occupations, for whom sideline activities are a supplementary activity of their spare time; and a minority of people who make their living entirely from sideline activities.

Sideline activities have long been an important part of economic life in the Chinese countryside. Most peasants engage in sideline activities of some sort during their spare time, either to supplement their income or to enrich their living. Sideline activities, therefore, are an essential part of the rural economy in spite of the fact that the value they create is much lower than, for instance, that of cropping.

Many of the sideline activities are carried out by individuals or households at scattered locations, and the value an individual or household creates may be quite small; it is difficult, therefore, to obtain precise statistics about the value created. The statistical data about sideline activities published by the State Statistical Bureau are inevitably incomplete.

In addition, the organization of production in the Chinese rural areas, and rural administration and management, have changed substantially since the end of the 1970s. In accordance with these changes, the meaning given to the term 'sideline' by the State Statistical Bureau has also been changed, and differs from the general definition given above. From 1958 to 1982, the State Statistical Bureau regarded all economic activities other than cropping, forestry, animal husbandry, and fishery, carried out by the basic units of the people's communes (that is by the production brigades or production teams) as rural sidelines. That implies that the rural sideline activities carried out by individuals or households were not included. It is true that few individuals or households were engaged in this type of production during the period mentioned, but that does not mean that there was no activity of this kind. Furthermore, the question is what the definition of rural sideline really was. For, on the other hand, various types of industries were regarded by the State Statistical Bureau as within the scope of rural sidelines at that time only if they were run by the production brigades or production teams. That means that all those industries run by brigades or teams, including rice-husking, wheat grinding, butchery, cotton ginning, mining, machinery manufacture, building materials manufacture, and even transport and construction businesses, were included in the scope of rural sidelines. The scale of these industrial activities, however, was very small, and they did not warrant being separated out into an independent category at that time.

Things have changed since 1982. Economic reform was started in 1979, and it was carried out first in the countryside. Thenceforth the Chinese rural economy developed rapidly. Not only did the industries run by the grass-roots units, regarded as within the scope of the rural sideline activities, develop substantially, but the sideline activities carried out by individuals or households — once regarded as 'tail of capitalism', which had been 'cut' several times — also flourished. In addition the new Constitution was published in 1982, and thereafter the nature of the people's commune, which was formerly an integration of both political and economic administrative functions, changed, and finally disintegrated. The basic units of the people's communes, the production brigades and production teams, also became villages or were even abolished, and are no longer basic units as before. The meaning of 'sideline', as previously defined by the State Statistical Bureau, then had to be changed.

The most important difference between the old and new definitions of rural sideline activities is that those industrial activities originally run by the production brigades and production teams are no longer regarded as within the scope of rural sideline activities. Instead, they are separated out into an independent sector called township enterprises. Thus, since 1984, according to the State Statistical Bureau's new definition, sideline activities consist of three economic activities: gathering wild plants; hunting wild animals; and the value of that part of the handicraft products which are commodity in nature (that is, for sale) and produced part-time by individuals or households. Thus, in comparison with the former definition, the new one includes the sideline activities of the peasant households and of individuals. It should be noted that only handicrafts which are produced as the result of part-time activities, and of these, only those which are for sale, are included. This means that handicrafts produced for self consumption are not included, nor are products other than handicrafts. The value of these self-consumed handicrafts, and of the products other than handicrafts, are included under the cropping, animal husbandry, forestry, or fishery sectors, according to their nature, under the new stipulation. Where households or individuals do not undertake cropping as their main activity, and only produce handicrafts part-time, they are treated as if producing handicrafts or engaging in industrial activities were their main activity, and the value of their products is accounted for not under rural sideline activities, but under township enterprises.

As shown in Table 3.1, the value of the sideline production as a percentage of the total agricultural production increased from 4.4 per cent in 1952 to 9.1 per cent in 1975, and then further to 15.1 per cent in 1979 and 22.0 per cent in 1984. These percentages were calculated in accordance with the State Statistical Bureau definitions prior to 1984. Figures taken from another source give this percentage as 30.2 in 1985, on the same basis. It is therefore obvious that sideline activities have developed rapidly since 1978. On the other hand, if the new definition is used, the part of the production formally included in the sideline activities but now included under township enterprises was 11.7 per cent in 1978, leaving only 2.9 per cent under the new definition of sideline production; in 1980 these two percentages were, respectively, 11.2 and 3.9 per cent; in 1984, 17.0 and 5.0 per cent; and in 1985, 24.9 and 5.3 per cent. These figures indicate how large the contribution of the rural enterprises was, within the original definition of sideline production, and how rapid the development of these enterprises has been in the past few years. This increase is even more apparent when actual monetary values are considered (see Table 3.1).

There is no doubt that rural sideline activities as now defined, that is the gathering of wild plants, the hunting of wild animals, and handicraft activities, are important and will continue to develop as a profitable occupation for surplus labour and for the use of leisure time after the farm work has been done.

Township enterprises

The rise of the township enterprises

As mentioned above, township enterprises is a term which has been used by the State Statistical Bureau only since March 1984; before that these enterprises were included under rural sideline activities. The Bureau defines township enterprises as those enterprises run by organizations below county level, such as townships or villages; those cooperative enterprises jointly managed by several households; and the newly developed individual enterprises in the countryside. In other words, township enterprises are all the non-agricultural enterprises in the material production sectors carried on in the countryside except for those run by the state — plus, at present, any agricultural enterprises run by a township or village; but these are few in number.

There was only limited industry in China before the 1950s. At that time there were only five types of craftsmen and four kinds of workshops in the Chinese countryside. They were the blacksmiths, carpenters, bricklayers, masons, and bamboo strip craftsmen; and the flour mills, noodle making workshops, oil mills, and bean curd workshops. Only in special situations were there some small workshops of other kinds. After the popularization of the people's communes some enterprises emerged in the countryside, but these were to undergo a series of contractions and suspensions and they hardly developed at all until the early 1970s. In the period that followed, township enterprises gradually grew in the areas where the foundation of industry was better, such as in coastal regions and in the suburbs of the big cities. The best example was in the countryside in southern Jiangsu province. With the implementation of the new economic policy in the rural areas from 1979 onwards, township enterprises rapidly emerged. At present, they have developed to such an extent that they account for quite a large proportion of production. Of the different types of industrial activity included in the category of township enterprises, the rural machinery industry was ranked first by value of production in 1985, and it accounted for more than one-quarter of the total value of production of the township enterprises; then followed the building materials industry which accounted for 19 per cent of the total, the textile industry 13 per cent, and all the remaining industries individually accounted for less than 10 per cent.

Apart from the implementation of the new economic policy, there were other reasons for the rise of township enterprises in China during the late 1970s. Firstly, the problem of surplus labour started to become serious in the 1970s because of the relaxation in the implementation of birth control and the advancement of medical treatment and sanitation. As a result, the average life span became longer and the infant mortality rate was reduced. The problem of surplus labour then became crucial, and the development of township enterprises was a way of relieving this problem.

Secondly, during the 1970s major parts of the large-scale industries were reduced to a semi-paralysed state by the riots caused by the 'cultural revolution' and markets were very short of commodities. This situation offered a good opportunity for the emergence of township enterprises to supplement market supplies.

Thirdly, agricultural mechanization had been strongly emphasized in the 1960s and 1970s, and many agricultural machinery repairing factories and tractor stations had been set up in the rural areas during that period. Although most of them were small and lacked good equipment, they nevertheless provided a basis for later development. Some of these factories and stations started to cooperate with the large factories in such a way that the large factories provided the raw materials and design, and the small rural factories and stations did the machining work. Township enterprises therefore developed more rapidly in places near to the industrially developed districts.

Fourthly, in the 1970s a large number of retired veterans from the factories returned to their native villages, and most of them were engaged by local enterprises as technical advisors. They were skilled workers who had close relations with large factories, and this enabled them to help the township enterprises obtain raw materials and find markets for their produce.

From 1980 onwards there was a rapid increase in the number of township enterprises and, together with the state run enterprises, they formed the non-agricultural material production sectors (the industries, construction, transport, and commerce) of the Chinese countryside (see Table 6.9). (The category of township enterprises also includes any agricultural enterprises run by townships or villages, but there are few of these). The total Gross Output Value of the township enterprises increased by 344 per cent between 1979 and 1985 (at 1980 prices), and the total employment by 92 per cent, with an annual increase of 28.2 and 11.5 per cent respectively. This was a much higher rate of increase in both income and employment than that which occurred in agriculture during the same period.

TABLE 6.9   The Gross Output Value of each of the five rural material production sectors as a percentage of the Rural Total Product of Society (RTPS)[a] in 1980 and 1986

| Sector | Whole country | | Rural economic regions | | | | | |
| | | | Eastern | | Central | | Western | |
	1980 (%)	1986 (%)	1980 (%)	1986 (%)	1980 (%)	1986 (%)	1980 (%)	1986 (%)
Agriculture	68.9	53.1	61.2	44.0	74.3	62.2	80.3	68.5
Rural industry	19.5	31.5	26.4	40.6	14.1	22.0	10.1	17.2
Rural construction	6.4	7.8	6.7	8.3	6.7	7.5	5.1	6.7
Rural transport	1.7	3.3	1.7	2.8	1.8	4.0	1.3	3.3
Rural commerce	3.5	4.3	4.0	4.3	3.1	4.3	3.2	4.3
RTPS	100.0	100.0	100.0	100.0	100.0	100.0	100.0	100.0

[a] The Rural Total Product of Society includes the Gross Output Value of Agriculture of state farms but not the contribution of state-owned units to the Gross Output Value in any other sector (see Appendix 2).
Source: State Statistical Bureau (1987). *Rural Statistical Yearbook of China 1987*. Chinese Statistical Publishing House, Beijing. (In Chinese)

The modes of development and the effects of the township enterprises

The township enterprises have developed in three different basic modes in three different regions as follows:

The Southern Jiangsu mode

This mode of township enterprise development appeared in the southern part of Jiangsu province and on the Liaotong Peninsula, and also in some suburbs of the big cities. In this mode the enterprises are chiefly township-run or village-run, and only a few are household or joint-managed enterprises. They are generally located in places with favourable geographic conditions and good transport facilities, and they also have good relations with urban industries. They are under the direct leadership of local government. Enterprises of this mode are generally quite large in scale and their management level approaches that of urban industries.

The Wenzhou mode

The Wenzhou mode originated in Wenzhou Prefecture, Zhejiang province. Its main characteristics are that the enterprises are mainly household enterprises. There is a fine division of labour and close cooperation among the people involved, and this cooperation is carried through to form a complete production and marketing system which is aimed at satisfying a particular market. This mode of township enterprise development depends little on the big cities. Household industries producing particular products are developed, and specialized markets for these products are supplied over a wide area, thus showing the vitality of a dispersed pattern of production of this type.

The Northern Jiangsu mode

This mode originated in the northern part of Jiangsu province. In it township and village enterprises are the mainstay, combined with some household or joint-managed enterprises. Characteristically there is an alliance of township enterprises, and they depend chiefly on the development and utilization of local resources. They also develop those services or processing businesses which are needed by urban industries. The main aim of the enterprises in this mode is to increase their level of production. The household and joint-managed enterprises, with their pooled funds and self-management, are solely responsible for their own profit or loss, and thus directly benefit (or not) the peasants concerned — they are characterized by a low level of investment, low consumption of energy, low costs, and high profits.

At present the township enterprises in the more developed regions are progressing towards even higher levels of organization, by, for example, arranging horizontal links and group management. With the deepening of the reforms in the rural areas of China more diversified types of township enterprises will take shape, and perhaps eventually provide an example of the industrialization of rural areas for the rest of the world.

But what role have the township enterprises actually played during their growth in the Chinese countryside?

Firstly, the development of township enterprises has pushed forward the development of agriculture. In places where township enterprises developed comparatively early and quickly, they accumulated substantial funds which could be used for agricultural production. They allocated 38.5 per cent of their profits to support agricultural production in 1978, and these funds were used to buy farm machinery, to construct irrigation schemes, to aid poor production teams, and to undertake other similar activities. Although the funds which the township enterprises have allocated to support agricultural production have been reduced in recent years, the enterprises still have a substantial positive effect on the development of agriculture through the subsidies given to the peasants in their localities. It is common in relatively developed regions for a part of the profits gained by township enterprises to be used for various construction projects as well as for subsidies to the members of the community in order to check the differences in income between the people engaged in cropping and the people working in the factories. This measure is called 'taking from industry to subsidize agriculture'.

Secondly, these township enterprises absorbed much surplus labour in the countryside, and speeded up industrialization nationally. According to an article in the weekly journal *Look-Out*, no. 23, 8 June 1987, the number of workers employed in township enterprises reached 73×10^6 in 1986, an increase of 45×10^6 on the 1978 level, and a figure that exceeded the total employees of the state enterprises by some $3-10 \times 10^6$.

As a rule, both in the past and now, during the course of economic development large numbers of peasants have tried to rush into the cities. China, however, has such a large population that it is unrealistic to expect that the rural population can be reduced in this way. But with the development of the township enterprises in the rural areas and their absorption of part of the rural labour force, the problem of surplus rural labour is less pressing and there is less pressure on the cities to absorb more people from the countryside. Right or wrong, this is the only solution to the Chinese problem of surplus labour in the rural areas.

Thirdly, the development of township enterprises changed the old economic pattern of the Chinese rural areas which had depended solely on crop cultivation. Those enterprises involved in the machinery industry, construction, transport, and commercial activities in the rural areas have helped to increase the industrialization of the countryside. In some places the township enterprises are mainly devoted to developing and utilizing local resources, and to providing services for the local people, and in these places manufacture of building materials, food processing, and provision of sewing services are common. In places where transport facilities are more numerous and better, and there is a better economic base, the township enterprises have become an extension of large industries and provide essential support to them. More and more township enterprises are becoming workshops for the large industries in the urban areas, and are obtaining modern technologies from them in turn.

Fourthly, the development of the township enterprises has provided additional sources of income for the peasants. In 1985 the total wage of the employees of township enterprises was $47\,210 \times 10^6$ yuan, and the average wage per employee was 676.5 yuan (*China Agriculture Yearbook 1986*). This was comparable to the average net income per labourer in the peasant households surveyed in 1985, and there is no doubt that the development of the township enterprises has provided opportunities for the peasants to improve their standard of living.

In summary, the functions of the township enterprises can be shown as depicted in Fig. 6.1.

Rural non-agricultural enterprises and the three rural economic regions

The rural non-agricultural enterprises are all the non-agricultural enterprises in the rural areas except for those run by the state. This concept differs from that of the township enterprises in that the latter also include the few agricultural enterprises run by a township or village. Owing to the diversity of natural, historical, and social conditions, economic development is subject to regional differences. Generally speaking, the whole of China can be divided into three apparently different rural economic regions — the eastern rural economic region, which is the most developed, the central rural economic region, which is moderately developed, and the western rural economic region, which is rather backward (see Fig. 6.2 and Table 6.9).

These three rural economic regions will be described below, but it should perhaps be explained that they are divided on an overall basis without being correct in every detail. In Sichuan province, for instance, there are some places where conditions are even better than in the central region, but they

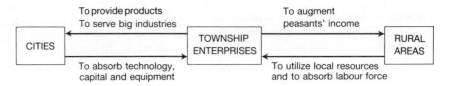

FIG. 6.1   The functions of township enterprises in China.

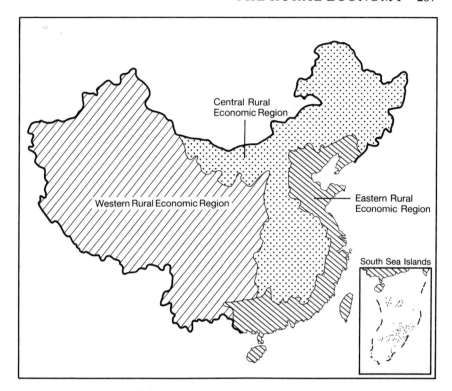

FIG. 6.2   The three rural economic regions of China.
Source: *China Daily*, 31 March 1986. Sketch map by Zhang Yaoning.

are included in the western region in view of the overall consideration as well as of the geographic factors. Similarly, not all the places in the central region are at the same level economically. Nevertheless, if this is kept in mind, the general overview should be a help in understanding the economic conditions of rural China.

The Eastern Rural Economic Region

This covers 11 provinces, autonomous regions, and municipalities, and 66 per cent of the gross output, by value, of the rural non-agricultural enterprises of China comes from this economic region. This rural industrial development is reflected in the movement of a considerable proportion of the labour force from the land. In 1985, 27 per cent of the total rural labour force was engaged in rural non-agricultural enterprises.

This development has proceeded to such an extent that the production of the rural non-agricultural enterprises now exceeds that of agriculture. In 1986 the production of the rural non-agricultural enterprises amounted to 56 per cent of the Rural Total Product of Society (RTPS) of the region; and the RTPS of the region amounted to 55 per cent of the RTPS of the whole country.

The Central Rural Economic Region

This region covers nine provinces and autonomous regions, most of which include extensive plains. Much of the land is fertile, and there are many good irrigation facilities, which help to make the region the production base for commodity 'grain'. During the period 1980–85 the Chinese government purchased nearly 200×10^6 tonnes of 'grain' from this region — half the total purchase of 'grain' in that period. In 1986 the production of the rural non-agricultural enterprises was 38 per cent of the RTPS of the region. In this economic region the old economic pattern, with crop cultivation as the main activity, remains. Nevertheless, 24 per cent of the gross output, by value, of the rural non-agricultural enterprises of China comes from this region.

The Western Rural Economic Region

This region includes nine provinces and autonomous regions. Most of the region is mountainous, desert, or consists of the deep gullies of the Loess Plateau. In general the natural conditions are not very good. The level of agricultural production is rather low, and the people living in the region tend to be poor. The rural non-agricultural enterprises are just beginning in this region, though in 1986 their production still accounted for 32 per cent of the RTPS of the region. About 10 per cent of the gross output, by value, of the rural non-agricultural enterprises of China comes from this region.

Land, labour, and capital in the Chinese rural economy

Land resources and land productivity

According to the data published by the State Statistical Bureau there are 96.85×10^6 ha of cultivated land in China. However, the figure obtained by remote sensing from a US satellite is 150.67×10^6 ha. The Chinese government is now checking this latter figure, but for the present we must use the figure given by the State Statistical Bureau. Due to varying natural conditions, parts of the cultivated land are subject to natural disasters. There are, for example, 4×10^6 ha of low lying land which are easily waterlogged, 6.67×10^6 ha of land which have a salinity problem, and further large areas of cultivated land which suffer from soil erosion. The effective irrigated area is about 44×10^6 ha, which is 45.5 per cent of the cultivated land. This figure is rather high compared with many other countries. The cultivated land of China occupies about one-tenth of the total territory of the country.

The area of cultivated land was increased substantially during the six centuries before 1949, as people successively reclaimed waste land; but now there is only limited waste land available that can still be reclaimed for cultivation, and, furthermore, a continuous loss of cultivated land. In 1952 the area of cultivated land was recorded as 108×10^6 ha and by 1978 this had fallen to 99.3×10^6 ha. This means that the area of cultivated land decreased by an average of 0.333×10^6 ha/year over this 26-year period. In recent years the rate of loss of cultivated land has been even greater, due to rapid progress in expanding the infrastructure in both urban and rural areas, and to increasing building in rural areas. It has been reported that the loss of cultivated land was 2.667×10^6 ha during the period 1978–85, an average of 0.381×10^6 ha/year (*People's Daily*, 17 February 1987). The government is already aware of this problem and is taking measures to limit the abuse of cultivated land.

The distribution of cultivated land in China is extremely uneven, with 92 per cent of it located in the humid and semi-humid monsoon regions in the east and south-east of China. It is this uneven distribution of cultivated land that has caused the uneven distribution of the population. Some 95 per cent of the population lives in the east and south-east of the country and this has resulted in great differences in the economic development of the different regions.

To cope with the problem of the rapid growth of population in the 1960s the government took measures to put more investment into agriculture and to intensify cropping, in order to increase the productivity of the cultivated land. As a result yields of the main crops were increased, as shown in Table 6.10.

The yields per hectare sown in present day China are generally a little higher than the world average, but lower than in the high-yielding countries. However, the big differences in China between the yields of the fertile and non-fertile lands, sometimes as much as four-fold, suggest that there is still potential for increasing the yields; there is no doubt that the principal aim for China's agriculture is to increase the productivity of the cultivated land.

Closely related land resources are the grasslands and the forest areas. The grasslands extend over 320×10^6 ha in the north and north-western part of China, and 70 per cent of this area is usable. The forests cover 115.24×10^6 ha, which is only 12 per cent of China's territory. The natural forests are concentrated in the north-eastern and south-western regions of China where over half of the forest area is to be found and three-quarters of the forest resource.

In China the forests have been over-felled, and the grasslands over-grazed to variable extents, and until these problems are resolved they will continue to result in deforestation and desertification. The area of desert in China has

TABLE 6.10   The average yield per hectare sown, of the major farm crops, in 1957, 1978, and 1985

	1957 (kg/ha)	1978 (kg/ha)	1985 (kg/ha)
Rice	2693	3975	5250
Wheat	855	1845	2940
Maize	1433	2805	3600
Soya bean	788	1058	1365
Cotton	285	443	810
Peanuts	1013	1343	2010
Rapeseed	383	720	1245

Source: Editorial Board of China Agriculture Yearbook (1986). *China Agriculture Yearbook 1985 (English Edition)*. Agricultural Publishing House, Beijing; Editorial Board of China Agriculture Yearbook (1987). *China Agriculture Yearbook 1986 (English Edition)*. Agricultural Publishing House, Beijing.

increased by nearly 40 000 km^2 in the last 25 years. The government has been calling on the people to launch a movement for afforestation and grassland sowing to make the country green, but it will take some time for trees to grow and for herbage to become established.

The third major resource is the fresh water. There are 16.64×10^6 ha of fresh water in China, of which about 5×10^6 ha can be used for aquaculture. Of this potentially usable area, 3.05×10^6 ha, or three-fifths, has actually been used. China has the largest number of reservoirs in the world and the area of reservoir water utilized for fish raising ranks third in the world after the USSR and USA. There are 85 000 reservoirs in China with an area of 2×10^6 ha suitable for fish raising, which is 40 per cent of the total area of fresh water suitable for this purpose.

The total catch of fish from the reservoirs was 206 000 tonnes in 1985. However, the reservoir area is not yet fully utilized, and the area that is utilized could be farmed more intensively. Furthermore the contribution of fishery production from reservoirs to total freshwater fish production was 20 per cent in 1980 but had dropped to only 8.7 per cent by 1985. This relative decrease reflects the superiority of the household contracted responsibility system, for it was the rapid increase in yield from the fish ponds managed by the households that raised fish production from this source at a much faster rate than that from the reservoirs.

As a rough estimate, in China the production value of 1 ha of crop land is equal to that of 12 ha of forest land, or 50 ha of grassland, or 3 ha of fresh water. The crop land now available is the mainstay of Chinese agriculture, providing three-quarters of the productive resource and producing 90 per cent of the primary products. Apart from the crop land, all of the forest land, grassland, and fresh water together provide only one-quarter of the productive resource, and produce only 10 per cent of the primary products. However, these latter three resources are very important for the ecological environment.

Labour resources and labour productivity

The labour resources in the rural areas are extremely abundant. There were some 304×10^6 people engaged in crop cultivation, animal husbandry, fishery, sideline activities, and forestry in 1985, which was about three-fifths of the total national labour force, and a further 67×10^6 people were engaged in rural non-agricultural activities, thus making a total number of 371×10^6 working people (labourers) in the rural areas (see Table 6.11).

Labour productivity in Chinese agriculture is lower than that in many other countries. In 1985 each agricultural labourer provided most of the food for an average of 3.44 persons in the total population of China. To do this he or she worked on 0.32 ha of cultivated land to produce 1249 kg of 'grain' and 58 kg of meat (see Table 6.12).

There are many reasons for the low labour productivity in Chinese agriculture, but the main reason is the increase in the population, which was accompanied by a decrease, rather than an increase, in the area of cultivated land. The area of crop land cultivated by each agricultural labourer was 0.62 ha in 1952, and it dropped to nearly half that by 1978, and dropped further still by 1985. In the 1960s and 1970s the Chinese government, besides developing the machinery and petroleum industries, also encouraged the improvement of agriculture through the introduction of high quality crop hybrids and varieties, through the application of more fertilizers and pesticides, and by increasing and improving irrigation facilities. Although this resulted in quite a large increase in crop yields, the productivity of the agricultural labourers increased only very slowly. The explanation for this is that, apart from the natural growth of the population within the rural areas, a large number of people moved into the rural areas from the urban areas. For example, many high school students were sent into the countryside to learn from the labourers there. In addition, the non-agricultural sectors in the rural areas were not able to absorb more of the labour force.

We will now consider the change in productivity of agricultural labourers in terms of the Net Output Value of Agriculture (NOVA) after the influences of the fluctuations in the purchase prices of agricultural products have been eliminated. Table 6.13 shows that during the 26 years from 1952 to 1978 the productivity of the agricultural labourers increased by only 3.2 per cent. After 1978, the adoption of the household contracted responsibility system linking income with output, and the raising of the purchase prices for agricultural products by the government, enhanced the peasants' enthusiasm for production and resulted in a large increase in the output of agricultural produce. Meanwhile, the restrictions on the peasants entering the cities was

TABLE 6.11 The rural labour force, 31 December 1985

Of the townships:	10^6
Total population	844.20
Number of households	190.77
Labour force	370.65
Rural non-agricultural labour force:	
Industry	27.41
Construction	11.30
Transport, posts and telecommunications	4.34
Domestic trade and catering trade	4.63
Property management, public utilities, inhabitant services, and consultancy services	0.89
Public health, sports, and social welfare	1.22
Education, culture and arts, broadcasting and TV	3.10
Scientific research and general technical services	0.13
Banking and insurance	0.11
Rural economic organizations	0.81
Others	13.19
Total rural non-agricultural labour force	67.13
Labour force in cropping, animal husbandry, sideline activities, fishery, and forestry	303.52
Total rural labour force	370.65

Note: The concept of 'rural' used in this table differs from that in Table 2.3, in that it includes the rural population living within the designated boundaries of the cities and towns. A description of how the rural statistics are collected will be found in Yu Min (1988). China's rural socio-economic statistical system. In Heung Keun Oh (Ed.) *Development of Food & Agricultural Statistics in Asia and Pacific Region, 1965–1987*. Korea Rural Economics Institute, Seoul. D49/1988.7. pp. 277–90.

In addition to the above labour force, there were 6.43 million Staff and Workers on state farms in agriculture, animal husbandry, fishery, and forestry. These people are not usually included in the totals of the 'rural' population or 'rural' labour force, but are kept quite separate in the national statistics.

Source: State Statistical Bureau (1986). *China: A Statistical Survey in 1986*. New World Press, and China Statistical Information and Consultancy Service Centre, Beijing; State Statistical Bureau (1987). *Yearbook of Rural Social and Economic Statistics of China 1986*. China Reconstructs, and China Statistical Information and Consultancy Service Centre, Beijing. (In English)

eased, especially in more recent years, and peasants were encouraged to leave the land and become active in the non-agricultural sectors. As a result, during 1978–85 the productivity of the agricultural labourers increased much more rapidly, at about 5 per cent per year. However, the productivity of the agricultural labourers is still low and further improvements will be subject to

TABLE 6.12   Labour productivity per year in Chinese agriculture, 1952–85

	1952	1965	1978	1985
Total agricultural labour force (10^6)	173.17	233.98	293.95	303.52
Per head of the agricultural labour force:				
No. of people in the total population	3.31	3.10	3.37	3.44
Area of cultivated land (ha)	0.62	0.44	0.34	0.32
'Grain' produced (kg)	958	831	1037	1249
Meat produced (kg)	20	24	29	58

Source: Department of Rural Statistics, State Statistical Bureau (1984). *The Glorious Achievements of Chinese Agriculture*. Chinese Statistical Publishing House, Beijing; Year 2000 Research Group (1984). *China in the Year 2000*. Science and Technology Literature Publishing House, Beijing; State Statistical Bureau (1986). *Extracts of the Statistics of China 1986*. Chinese Statistical Publishing House, Beijing; *The People's Daily*, 17 February 1987; State Statistical Bureau (1986). *Rural Statistical Yearbook of China 1986*. Chinese Statistical Publishing House, Beijing. (In Chinese)

TABLE 6.13   Agricultural labour productivity per year as calculated from the Net Output Value of Agriculture (NOVA), 1952–85

	1952	1957	1965	1978	1985
Index of purchase prices for agricultural products	100	120.2	154.5	178.8	298.4
Net Output Value of Agriculture (NOVA) (10^8 yuan)	340	425	641	1065	2492
Agricultural labour force (10^6)	173.17	193.10	233.98	293.95	303.52
NOVA per head of labour force (yuan/person)	196.34	220.09	273.96	362.31	821.03
NOVA per head of labour force at constant 1952 prices (yuan/head)	196.34	183.10	177.32	202.63	275.14

Source: State Statistical Bureau (1986). *China Statistics Yearbook 1986*. Chinese Statistical Publishing House, Beijing. (In Chinese)

two factors. The first is whether the non-agricultural sectors can provide further employment opportunities. The second is whether more capital can be made available for agricultural production in order to bring about more capital intensive farming.

*Capital substitution and the ratio of production expenditure to
income*

Chinese agriculture is labour intensive and meticulous, but capital investment per hectare of cultivated land is very low. According to the survey of
peasant households undertaken by the State Statistical Bureau at the end of
1985, the productive assets of each household were, on average, worth 793
yuan. Land is the property of the collective and therefore is not included as a
property of the household, but the value of the capital assets of the Chinese
peasants are still low per labourer when compared to the assets other than
land enjoyed by farmers in other countries. The value of the items used in
production, such as fertilizers and pesticides, when apportioned per unit area
of cultivated land is also quite low. As many scholars have pointed out, there
has to be an increased input of capital if the long term productivity of
agriculture is to be increased.

In substituting capital for labour in order to raise land and labour
productivity, China is confronted with two obstacles. One is that the labour
force liberated through substitution can not be fully utilized and this then
leads to major unemployment. The other is that the prices of agricultural
products are too low to induce the peasants to spend money on crop production, and as long as this situation prevails then increasing the capital
intensiveness of crop production will not be possible.

While capital substitution is encountering serious obstacles in the cropping
sector, it is advancing quite well in other sectors. The most obvious example
is in the increase in vehicles used for transport. Modern means of transport,
such as trucks and tractors, together with old carts fitted with rubber tyres,
have all become favourite purchases of the peasants in recent years in the
rural areas. It is probably not necessary to add that the development of
transport naturally offers more opportunities for employment. Although the
number of tractors has increased rapidly, as previously indicated, their main
use is not in agriculture to increase labour productivity, but in the non-
agricultural sectors, especially transport, where tractors have increased both
employment opportunities and income. According to a survey of 1 178 700
households in rural areas in 1985, the income from the operation of tractors
was 7467×10^6 yuan, of which that from transport was 5577×10^6 yuan, or
74.6 per cent of the total income from using tractors.

From the point of view of the development of different sectors in China's
rural areas, capital substitution has already occurred in the non-agricultural
sectors, and it is also appearing in intensive farming in places which are
relatively well developed economically, and/or where there is a greater
availability of cultivated land per caput. As the cash income of rural
households increases (as compared with production for self-consumption
and income in kind), the tendency to purchase more capital assets is

beginning to appear in the rural areas as a whole. The survey of peasant households at the end of 1987 revealed an average expenditure of 3.13 per cent of calculated gross income on capital items for use in production (see Table 6.14).

The costs, prices, and subsidies of agricultural products

Owing to regional differences and the uneven development of the economy, the costs of agricultural products differ greatly in different places. The cost of labour is rarely taken into account by rural households in their productive activities, and the aim of production in the rural areas is not to obtain the highest profit, but to obtain the largest disposable income. Only the astute specialized households consider the cost of labour because they are engaged in commodity production, and they need high labour productivity to enhance their competitive advantage. Furthermore, the specialized households sometimes hire one or two farm hands to help them, and they tend to consider whether incurring the cost of the additional labour will result in a higher net income. The cost of labour varies very much between different regions in China. In the poor regions the cost is very low, less than one yuan per working day on average, while in the east it is much higher and in some cases reaches 10 yuan per working day. The costs and returns for different crops also vary regionally and, because these regional differences are very large, it is difficult to obtain precise statistics about the costs of crop production. It is, however, evident from the available statistics that the costs and returns of cash crops per unit area are substantially greater than those of 'grain' crops (see Table 6.15).

In recent years the costs of agricultural production have risen rather rapidly because of adjustments in the prices of various items used in production. In addition, total national 'grain' production dropped considerably in 1985, and this detracted from any increase in income for the peasants who were engaged mainly in 'grain' production. Such events have tended to make people pay attention to the costs of increasing agricultural production. An article in the weekly journal *Look-Out*, no. 21, 25 May 1987, reported details of a rural household on a plain in central China, which cultivated 10.5 mu (0.70 ha) of contracted land. The average expenditure on fertilizer, diesel oil, machinery-tillage fee, and seed, amounted to 80 yuan per mu in 1985. The expenditure on diesel oil alone was as much as 15 yuan, or 18 per cent of current expenditure. The yield of wheat was generally 250 kg/mu (3730 kg/ha), and the net income for wheat was less than 60 yuan. If the income from the autumn harvest 'grain' crops is added to this then the total net income comes to just a little more than 100 yuan per mu annually.

Under the present 'double track' policy, there are two sets of prices for the agricultural means of production; the favoured or common prices, and the

TABLE 6.14 The average annual income and expenditure of peasant households, based on household surveys, 1981–87

	1981	1982	1983	1984	1985	1986	1987
Number of households surveyed	18 529	22 775	30 427	31 375	66 642	66 836	66 912
Average number of permanent residents per household	5.50	5.46	5.43	5.37	5.12	5.07	5.01
Average number in the labour force per household	2.53	2.58	2.84	2.87	2.95	2.95	2.95
Average income and expenditure in yuan per permanent household resident:							
Calculated gross income	NA	NA	424.81	485.38	556.12	592.94	656.35
Less production costs and taxes:							
Household production expenses	NA	NA	80.28	95.67	121.39	132.65	150.59
Capital items for production	6.27	11.69	18.44	16.89	18.70	16.66	20.52
Reponsibilities contracted to collectives	NA	NA	10.35	10.92	10.79	11.80	13.85
Taxes	NA	NA	5.97	6.57	7.64	8.07	8.84
Total production costs and taxes	NA	NA	115.04	130.05	158.52	169.18	193.80
Net income	223.44	270.11	309.77	355.33	397.60	423.76	462.55
Living expenditure	190.81	220.23	248.29	273.80	317.42	356.95	398.29
Remittances to family members, gifts to relatives, and other expenditure	12.59	14.04	17.14	17.85	22.21	9.69*	11.90*

TABLE 6.14   (Cont.)

	1981	1982	1983	1984	1985	1986	1987
Balance after deducting living and other private expenditure	20.04	35.84	44.34	63.68	57.97	57.12	52.36
Cash in hand plus bank balance at year end	35.76	49.71	60.24	92.42	112.55	149.16	191.74

*Only includes gifts to relatives outside the rural areas.

NA = Not available. Data was either not collected or is unreliable.

Note: The 'Calculated gross income' was obtained by adding together the 'Total production costs and taxes' and the 'Net income'.

The 'Capital items for production' are usually called 'productive fixed assets' in English language Chinese published statistics; they include houses and buildings used for productive purposes; draught animals and product animals; machinery for agriculture, forestry, animal husbandry, and fishery; machinery for industry and sideline occupations; and vehicles for transportation.

Household labour costs in household production are not included, and it should be assumed that the above figures represent value both in cash and in kind, and include the value of items produced by a household for its own consumption.

Source: State Statistical Bureau (1986). *Rural Statistical Yearbook of China 1985*. Chinese Statistical Publishing House, Beijing; State Statistical Bureau (1986). *Rural Statistical Yearbook of China 1986*. Chinese Statistical Publishing House, Beijing; State Statistical Bureau (1988). *Rural Statistical Yearbook of China 1988*. Chinese Statistical Publishing House, Beijing. (In Chinese)

TABLE 6.15   The production costs and profit of growing rice, wheat, maize, cotton, peanuts, and sugar cane, and of producing hogs, in 1984

Crops	Rice	Wheat	Maize	Cotton	Peanuts	Sugar cane
Gross output value[1] (yuan/mu)	142.6	99.0	99.5	249.0	143.3	367.6
Material production cost (yuan/mu)	43.2	36.4	31.8	58.3	42.8	118.6
Net output value (yuan/mu)	99.4	62.6	67.7	190.7	100.5	249.0
Labour cost[2] (yuan/mu)	34.0	23.7	25.3	68.5	35.1	90.7
Taxes (yuan/mu)	3.2	2.4	2.2	4.3	2.4	7.1
Net profit (yuan/mu)	62.2	36.5	40.2	117.9	63.0	151.2
Labour required (man-days/mu)	22.7	15.8	16.9	45.7	23.4	56.2
Net output value per man-day of labour (yuan)	4.4	4.0	4.0	4.2	4.3	4.4

Hogs raised by households

Sale price (yuan/hog)	166.9
Estimated value of manure, as a hog by-product (yuan/hog)	20.6
Material production cost (yuan/hog)	133.6
Net output value (yuan/hog)	53.9
Labour cost[2] (yuan/hog)	38.1
Net profit (yuan/hog)	15.8
Labour required (man-days per hog)	25.2
Net output value per man-day of labour (yuan)	2.1

1 mu = 0.0667 ha or 0.1647 acres

[1]Including the value of by-products.

[2]The estimated cost of labour is 1.5 yuan per man-day (rounded to the nearest 0.1 yuan) for the production of all the crops except sugar-cane, and for the production of hogs; the estimate for sugar-cane production is 1.6 yuan per man-day.

Source: State Statistical Bureau (1987). Rural Statistical Yearbook of China 1986. Chinese Statistical Publishing House, Beijing. (In Chinese)

negotiated prices. The former are set by the government, and the latter are controlled by the free market mechanism. If conditions are normal, peasants are able to obtain their means of production at the favoured prices. However, quite often something happens and the peasants can not buy enough of certain items at the favoured prices, and they then have to buy at negotiated prices; in most cases these are 50 per cent, and may be even several times, higher than the favoured prices. For example, in the rural household mentioned above, 10 kg of diesel oil had to be bought to irrigate the crops, but only 1 kg could be bought at the favoured price and the remaining 9 kg had to be bought at the negotiated price.

It is worth noting here a few special points concerning costs in China. The price of land is not included when determining the costs of agricultural products, because of the public ownership of the land. Also, as mentioned above, in most cases the peasants do not consider the cost of labour, and the estimates concerning labour in the official calculations are generally too low, as it is difficult to represent the universal conditions. Hence the unified prices for agricultural products, set according to cost calculations by the government, are inclined to be too low. In the more developed eastern part of China, the non-agricultural sectors are advancing faster than in the other regions, and the price of labour here is relatively high; as a result there is an obvious lack of incentive for 'grain' production. Another well recognized phenomenon is that, from the point of view of an input–output ratio, to grow 'grain' crops is less favourable than to grow cash crops, and growing cash crops is less favourable than working in a factory or at a trade. Generally the input–output ratio for industry is 1:8; for a trade or service business 1:6; for cash crops 1:5; and for 'grain' crops 1:3 (*China's Rural Economy*, February 1987, p. 29). This is another factor which influences the enthusiasm of peasants for growing 'grain' crops.

At present there are three types of prices for agricultural products in China:

1. planned prices, set by the government, including the buying prices for agricultural products set by different levels of government;

2. fluctuating prices — these are planned prices which fluctuate within a range, the range of fluctuation and the commodities covered are determined by the department concerned;

3. free prices, which are mutually agreed upon by the buyer and the seller through negotiation.

In country fairs most of the bargains are made at free prices which fluctuate with the variations in supply and demand. The government controls the free prices through state trading activities in the market and through supply and marketing cooperatives.

During the years from the 1950s until 1985 China enforced strict direct supervision over prices: this was called the system of unified purchase and marketing (and was similar to the practice in the USSR and Eastern European countries), and its essence was the state's monopoly of some agricultural products. The unified purchase prices were generally lower than the market prices, and it goes without saying that the difference was of the nature of additional taxation. On the other hand, the state selling prices of agricultural products were also usually lower than the market prices, and sometimes the state selling prices were even lower than the state purchase prices. The difference in such cases was a form of subsidy to the consumers of the agricultural products. As far as the state revenue was concerned, such subsidies were a large financial burden; in 1984 they accounted for 37×10^9 yuan, or one-quarter of the revenue of that year.

The state abolished the state purchase system in 1985 and instead adopted a system of 'contracted purchase', which was, however, still carried out at prices supervised by the government. Since 1978 the government has gradually relaxed control over prices. After it raised a wide range of purchase prices of agricultural products in 1979, the government narrowed the range of items to which unified purchase was applied, and correspondingly widened the range of agricultural products for which prices were supported by the government. In 1985 the government stipulated that after the contracted purchase had been fulfilled the remaining products could be disposed of freely, subject to the variations of the market mechanism. In particular, the government no longer exercised strict control over prices for meat, eggs, vegetables, and fruit. Because there is quite a large difference between the state purchase price and the market price, the peasants' enthusiasm for fulfilling their duty of selling definite amounts of agricultural products as stipulated in their contract is now dampened, and the government has had to increase again the subsidy for the means of production to remedy this. Hence the government has sustained a much heavier burden of such subsidy in 1986–87.

The life of the Chinese peasants, their income and expenditure

Under the present household contracted responsibility system, a peasant household, in general, has the following characteristics:

1. Management is small-scale. According to the 1985 survey of households in rural areas, on average, each household had 0.56 ha of contracted, cultivated land which was divided into 9.7 plots of 0.06 ha.

2. A peasant household is an independent and integrated economic entity, with self-determination in production.

3. The overwhelming majority of households grow mainly field crops, especially 'grain' crops. The produce is used for self-consumption, for feeding to farm livestock, and about 20 per cent of it is sold on the market.

4. Peasant households make only a small cash outlay on their production, and they receive only a limited cash income; they live mainly on crop cultivation and do not borrow much from others.

5. The area of cultivated land managed by a peasant household is related to the availability of resources and transport, and also to the development of the non-agricultural sectors, for the latter influence the possibility, and the extent, of the peasants leaving the land.

6. The peasant household is not only a unit of production management, but also a unit of consumption. The main source of income is cropping, but there is a part of the income which comes from the non-agricultural sectors.

The income of peasant households

The peasants' net incomes increased by 78 per cent during the period 1978–85 (after allowing for price inflation). Previously, from 1957 to 1978, the overall increase had only been 23 per cent on the same basis, and since 1985 the rate of increase has again slowed down (see Table 6.16, and Table 6.13 for inflation data).

The high rate of increase in income during 1978–85 was a special case, and it is difficult to find a comparable example either in Chinese history or in other countries. The reasons are various. First of all, there was the adjustment of production relations resulting from the government's rural reforms. The new household contracted responsibility system inspired the peasants to work to their best ability. Secondly, the government increased the prices of agricultural products in 1979 and replaced many controlled prices with supported prices; this encouraged higher levels of production which had not been realised in the past due to lack of incentives. After 1985 the situation became more normal, but the per caput net income of the peasants is still increasing more rapidly than it did in the years before the rural reforms were introduced.

Since the implementation of the household contracted responsibility system in the early 1980s, and the disintegration of the people's communes, the households in the rural areas have become independent management units, and their main source of income has changed from the collectively managed units and unified distribution of the past, to their own household. The average net income per head (of all people in the household, whether working or not) from household managed businesses in 1986 was 345.28 yuan, which was 81.5 per cent of the average total net income, whereas in the period 1957–78, a period of highly collectivized production, the percentage was only 30.7 per cent (see Table 6.16).

As the structure of the economic sectors in the rural areas changed from a simple structure to a composite structure with multiple levels, the sources of income of the rural households also changed. Firstly, the former policy of

TABLE 6.16 The sources of the net annual income of peasant households, 1954–86

	1954	1957	1965	1978	1985	1986
Net income per permanent resident (yuan)						
Net income from collectives	2.43	43.40	63.17	88.53	33.37	36.15
Net income from economic unions					3.69	2.92
Net income from household managed business	56.39	21.46	33.29	35.79	322.53	345.28
Other non-borrowed income	5.32	8.09	10.74	9.25	38.01	39.41
Total income	64.14	72.95	107.20	133.57	397.60	423.76
Percentage of income from each source						
Net income from collectives	3.8	59.5	58.9	66.3	8.4	8.5
Net income from economic unions					0.9	0.7
Net income from the household managed business	87.9	29.4	31.1	26.8	81.1	81.5
Other non-borrowed income	8.3	11.1	10.0	6.9	9.6	9.3

Note: 'Net income from collectives' is the total income received from the collectives and includes the income from the basic accounting units, from other collective units of different levels, and the part of the income from contracted collective production that is private income.

'Other non-borrowed income' includes the income, both in cash and in kind, brought back by household members from outside the household, and poverty allowances, public work allowances, and disabled servicemen's allowances from the government, etc.

Source: Department of Rural Statistics, State Statistical Bureau (1984). *The Glorious Achievements of Chinese Agriculture*. Chinese Statistical Publishing House, Beijing, (In Chinese); Editorial Board of China Agriculture Yearbook (1988). *China Agriculture Yearbook 1987 (English Edition)*. Agricultural Publishing House, Beijing.

'taking "grain" as the key link', which emphasized only 'grain' production, has come to an end. In addition to growing 'grain' crops, peasants now also engage in multi-sector economic activities, or in diversified agriculture. Secondly, the development of the non-agricultural sectors in the rural areas, especially of the various township enterprises, has brought about new

TABLE 6.17   The composition of the net annual income of peasant households in 1978 and 1985

	1978		1985	
	(yuan/permanent resident)	(%)	(yuan/permanent resident)	(%)
Agricultural income	113.47	85.0	263.81	66.3
Non-agricultural income	9.39	7.0	86.26	21.7
Total production net income	122.86	92.0	350.07	88.0
Non-production income	10.71	8.0	47.53	12.0
Total net income	133.57	100.0	397.60	100.0

Note: 'Agricultural income' is the net income from crop cultivation, forestry, animal husbandry, sideline activities, and fishery.
'Non-agricultural income' is the net income from rural industries, construction, transportation, trade, and service businesses.
'Non-production income' includes the income, in cash and in kind, brought back by household members from outside the household, the income from the public welfare fund, accumulation fund, and finance from the state, etc.
Source: State Statistical Bureau (1986). *China Statistics Yearbook 1986*. Chinese Statistical Publishing House, Beijing. (In Chinese)

opportunities for the peasants to increase their income: 85 per cent of their income was from agriculture in 1978, but this figure had dropped to 66 per cent by 1985 (see Table 6.17).

Furthermore, commodity production has progressed rather quickly in recent years. According to a survey of 36 667 peasant households in 1984, income from selling products, processing, and providing labour was 59 per cent of the total income — a proportion quite different from that of former times. In the past the peasant's income was mainly an income in kind, they had very few things to sell. The increase in the marketing of agricultural products, and the increase in income from the non-agricultural sectors, reflect the rise of commodity trading within the rural economy. In turn, this increase in cash income has made possible the increase in cash expenditure shown below in Table 6.24.

The distribution of income

One of the aspects of the rural reforms is to allow and encourage some of the peasants to obtain greater profits than others. The general saying is 'Let a part of the people be better off first', or, 'The first profit makers bring along other profit makers'. The result of this policy has been to enlarge the differences in income between peasants, but it is believed that this is not yet bringing about a polarization. The general tendency is for the proportion of high-income people to grow, and of low-income people to fall. The present rural policy and reforms have alleviated poverty in the countryside. But there were still 5.1 per cent of peasants in 1984 with a per caput net income of less than 150 yuan. These people were extremely poor, and they still have difficulty subsisting.

The Gini coefficient of the Chinese peasants' income was 0.237 in 1978, and 0.264 in 1984. This rise was due to the breaking up of unified distribution — the so-called 'eating from the big pot' — and the consequent closer relationship between income and labour. Even so, Chinese society in the rural areas at present is still more or less a society of equality, for the Gini coefficient is near 0.2, the lowest value estimated by the United Nations for any country. This is also reflected in the percentage distribution of the incomes of the highest and lowest income groups. It is reported, that in 1984 the 20 per cent of peasants with the lowest income received 10.12 per cent of the total peasant income, whereas the 20 per cent of peasants with the highest income received 36.00 per cent (see Fig. 6.3).

Generally speaking, implementing the policy of 'making a part of the people better off first' has enabled a majority of the peasants to have an improved standard of living.

There exists in the Chinese rural areas a phenomenon which may be called 'imitation effect' due to the influence of traditional community culture. If

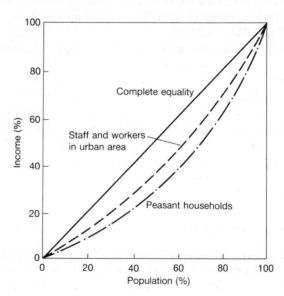

FIG. 6.3   The Lorenz curves of income distribution of Chinese peasant households, and of staff and workers in urban areas, in 1984.

one household engages in a particular type of production, and becomes a better off specialized household, then a very strong imitative psychology will appear amongst the neighbours, and this in turn will result in two households, three households, 10 households, and even most of the peasants in the village, coming to engage in that type of production. If there is a sufficient market for that commodity then all the peasants of the village will have more income, and a rich village will come into being. In the last few years some rich villages have emerged. However, as the level of technology involved rises, the types of production which can be imitated in this way will grow less, and this will, of course, increase the differences in income between the villagers.

The household expenditure of the peasants

The Chinese spend more than half of their total income on living expenses. Generally speaking, the consumption level of the peasants is rather low, but in recent years the growth in their income has brought a corresponding increase in consumption.

In 1978 the average annual raw 'grain' consumption of the peasants was 248 kg per person, and this had increased to 259 kg by 1987 (raw 'grain' is 'grain' after threshing but before any further processing such as rice and millet husking or wheat grinding has been carried out). Although this was only a small increase in the overall amount of 'grain' consumed, the types of 'grain' consumed did nevertheless change. The percentage of rice and wheat (taken together) in total 'grain' consumption increased from 49.6 per cent in 1978 to 81.5 per cent in 1987. Chinese peasants consider rice and wheat to be better than other 'grains' and so these grains are referred to as 'fine grains'; all other 'grains' are termed 'coarse grains'. The per caput consumption of meat, cooking oil, eggs, and poultry also increased quite substantially. Meat consumption (including poultry meat), for example, increased from 6.01 kg per person per annum in 1978 to 12.80 kg in 1987. According to the nutritionists, the diet of the Chinese peasants provides sufficient energy to meet their requirements, but it is deficient in protein and fat (see Tables 6.18 and 6.19).

It is worth noting that the Chinese peasants' social activities are conducted at the meal table, and that dietary customs formed over the long history of the Chinese peoples still influence food consumption in the countryside in China. The Chinese peasants live frugally in ordinary times, but on wedding days and at the time of funerals they are lavish with food in order to entertain relatives and friends. Also on festival days, or on the birthdays of family members, the diet is improved and an adjustment made to the ordinary simple living, but the low income of the peasants makes it difficult for them to obtain more animal protein.

The Chinese peasants have traditionally worn mainly cotton clothing, but

TABLE 6.18 The annual consumption of foodstuffs per person in peasant households, 1978–87

Item	1978	1983	1985	1986	1987	Increase, 1978 to 1987
	(kg)	(kg)	(kg)	(kg)	(kg)	(%)
'Grain' (raw)	248	260	257	259	259	4
Rice and wheat	123	196	209	212	211	72
'Grain' other than rice and wheat	125	64	48	47	48	– 62
Vegetables	142	131	131	134	130	– 9
Cooking oil	1.96	3.53	4.04	4.19	4.69	139
Pork, beef and mutton	5.76	9.97	10.97	11.79	11.65	102
Poultry	0.25	0.82	1.03	1.14	1.15	360
Eggs	0.80	1.57	2.03	2.08	2.25	181
Aquatic products	0.84	1.59	1.64	1.87	1.96	133
Sugar	0.73	1.26	1.46	1.59	1.70	133
Alcoholic drinks	1.22	3.20	4.37	4.96	5.48	349

Note: The figures in this table are from the surveys shown in Table 6.14, whereas the figures given in Table 6.1 are the averages calculated for the population of the whole country.
Source: State Statistical Bureau (1986). *China Statistics Yearbook 1986*. Chinese Statistical Publishing House, Beijing; State Statistical Bureau (1988). *Rural Statistical Yearbook of China 1988*. Chinese Statistical Publishing House, Beijing. (In Chinese)

TABLE 6.19 The daily nutritional intake of the peasants in 1978 and 1983

	1978	1983
Energy (kcal/person)	2224.0	2805.9
Protein (g/person)	68.5	81.7
Fat (g/person)	25.7	40.7

Source: State Statistical Bureau (1984). *China Statistics Yearbook 1984*. Chinese Statistical Publishing House, Beijing. (In Chinese)

in the 1980s, as the prices of chemical fibres have fallen and the peasants' incomes have grown, the consumption of synthetic fabrics has rapidly increased to exceed that of cotton cloth. Any improvement in the peasants' clothing is seen first of all in the clothing of the young people who are most easily influenced by the urban atmosphere. For weddings both families will spend much money on buying high-quality clothes and on some durable goods for living (see Table 6.20).

The peasants buy the consumer goods which they can not produce for themselves. They are mainly self-sufficient in food and fuel, but clothing, household utensils, and other consumer goods have to be bought from the markets. The proportion of the peasants' overall home consumption which was bought rather than self-produced increased from 39.7 per cent in 1978 to 64.5 per cent in 1987. Among these purchases were some items of a durable nature produced by industry. Although the numbers of durable items, such as television sets, purchased by the peasants are still low, they have been increasing very rapidly in recent years as shown in Table 6.21. When it is considered how large the population of the Chinese rural areas is, it is quite clear that the capacity of the Chinese rural market is enormous.

The structure of household expenditure

Table 6.22 shows that there has been a considerable change in the structure of the peasants' expenditure since 1978. Because the peasants have to undertake house building and take care of their expenditure on medicines themselves (in urban areas, the government spends much money on these two items as a subsidy), the increase in expenditure on these two items between 1978 and 1987 has exceeded that for other items. On the other hand, the percentage of the expenditure spent on food decreased from 68 per cent in 1978 to 55 per cent in 1987, although the actual amount spent still increased even after allowing for inflation.

The Engel coefficient (= expenditure on food/total expenditure × 100) in Chinese rural areas was 58 per cent in 1985. According to some of the

TABLE 6.20 The annual consumption of clothing per person in peasant households, 1978–87

Item	1978	1980	1985	1986	1987	Increase, 1978 to 1987 (%)
Cotton cloth (m)	5.63	4.30	2.54	1.98	1.57	– 72
Cotton (kg)	0.40	0.38	0.43	0.34	0.52	30
Synthetic fibre cloth (m)	0.41	0.94	2.50	2.53	2.28	456
Woolen cloth (m)	0.02	0.06	0.14	0.13	0.12	500
Silk and satin (m)	0.02	0.06	0.07	0.06	0.05	150
Knitting wool and woollens (kg)	0.02	0.05	0.04	0.05	0.07	250
Rubber, gym and leather shoes (pair)	0.32	0.51	0.55	0.66	0.68	113

Source: State Statistical Bureau (1986). *China Statistics Yearbook 1986*. Chinese Statistical Publishing House, Beijing; State Statistical Bureau (1988). *Rural Statistical Yearbook of China 1988*. Chinese Statistical Publishing House, Beijing. (In Chinese)

TABLE 6.21   The durable goods owned per hundred peasant households, 1978–87

Item	1978	1980	1985	1986	1987
Bicycles	30.73	36.87	80.64	90.31	98.52
Sewing machines	19.80	23.31	43.21	46.99	49.79
Radios	17.44	33.54	54.19	54.24	52.98
Clocks and watches	51.75	–	163.64	–	–
of which, watches	27.42	37.58	126.32	145.06	161.22
Television sets	–	0.39	11.74	17.28	24.38
Tape recorders	–	–	4.33	6.60	9.68
Washing machines	–	–	1.90	3.22	4.78
Electric fans	–	–	9.66	13.63	19.76
Refrigerators	–	–	0.06	0.20	0.31

Source: State Statistical Bureau (1986). *China Statistics Yearbook 1986*. Chinese Statistical Publishing House, Beijing; State Statistical Bureau (1988). *Rural Statistical Yearbook of China 1988*. Chinese Statistical Publishing House, Beijing. (In Chinese)

prevailing criteria used in other countries, the living standards of the Chinese peasants are still hardly at a subsistence level. Such a judgement, however, would not be in accordance with the actual situation. The reason is the difference in the meaning of the term food expenditure. In Table 6.22, food expenditure includes not only that expenditure necessary for subsistence, but also the expenditure made on banquets and gifts. Therefore the proportion of the total outlay on food is higher than normal, and this in turn suggests that application of the Engel coefficient is not appropriate. This argument may be taken a little further. Table 6.23 shows the result of a sampling investigation into the households of two different provinces. From this table it will be seen that the Engel coefficients for the two different regions are about the same, but the income levels differ quite substantially. The composition of the food in the two provinces is also different. While the expenditure on staple food ('grain' and beans) was less than that on non-staple food (meat, eggs, milk, and vegetables) in Zhejiang province, it was just the reverse in Shandong province. It is quite clear that the living standards of these two provinces are different, even although their Engel coefficients are about the same. This means that the Engel coefficient can not express exactly the quality of life, or the expenditure level, in the Chinese rural areas. However, it is true that the income level of Chinese rural households is rather low in terms of the composition of the diet that it can provide, and it is more or less similar to that of other low income countries.

Turning again to the structure of the living expenditure, the value of the commodities actually purchased, as compared to those produced and consumed by the household, increased from 39.7 per cent of the total value in 1978 to 64.5 per cent in 1987 (see Table 6.24). This shows that the Chinese

TABLE 6.22   The structure of the living expenditure of peasant households, 1978–87

Living expenditure	Expenditure per person						Distribution of expenditure					
	1978 (yuan)	1980 (yuan)	1981 (yuan)	1985 (yuan)	1986 (yuan)	1987 (yuan)	1978 (%)	1980 (%)	1981 (%)	1985 (%)	1986 (%)	1987 (%)
Food	78.6	100.2	113.8	183.3	201.2	219.7	68	62	60	58	56	55
Clothing	14.7	20.0	23.6	31.3	33.7	34.2	13	12	12	10	10	9
Fuel	8.3	9.7	10.6	18.2	18.6	19.3	7	6	6	6	5	5
Housing	3.7	12.8	18.7	39.4	51.2	57.8	3	8	10	12	14	14
Daily necessities, medicines, etc.	7.6	15.3	19.5	36.1	41.0	47.2	6	9	10	11	12	12
Total consumption expenditure	112.9	158.0	186.2	308.3	345.7	378.2	97	97	98	97	97	95
Cultural activities and services	3.2	4.3	4.6	9.1	11.2	20.2	3	3	2	3	3	5
Total living expenditure	116.1	162.3	190.8	317.4	356.9	398.4	100	100	100	100	100	100

Note: Expenditure figures are at prices of the particular year.
Source: State Statistical Bureau (1986). *China Statistics Yearbook 1986*. Chinese Statistical Publishing House, Beijing; State Statistical Bureau (1988). *Rural Statistical Yearbook of China 1988*. Chinese Statistical Publishing House, Beijing, (In Chinese); Editorial Board of China Agriculture Yearbook (1988). *China Agriculture Yearbook 1987 (English Edition)*. Agricultural Publishing House, Beijing.

TABLE 6.23 The net income, and the relative food expenditure per person, of peasant households in Zhejiang and Shandong provinces in 1985

Province	Net income (yuan)	Percentage of total living expenditure spent on food				
		Total food (%)	'Grain' & beans (%)	Meat, eggs & milk (%)	Other food (%)	Eating or drinking outside (%)
Zhejiang	548.60	52.15	17.15	23.18	10.24	1.58
Shandong	408.12	52.24	25.15	17.70	8.49	0.90

Source: Anon. (1987). *Statistics*, no. 1.

TABLE 6.24 The percentage self-sufficiency in living expenditure of peasant households, 1978–87

Item	Purchased				Self-supplied			
	1978 (%)	1980 (%)	1985 (%)	1987 (%)	1978 (%)	1980 (%)	1985 (%)	1987 (%)
Total living expenditure	39.7	50.4	60.2	64.5	60.3	49.6	39.8	35.5
Food	24.1	31.1	41.7	46.4	75.9	68.9	58.3	53.6
Clothing	89.0	98.1	98.0	98.0	11.0	1.9	2.0	2.0
Housing	95.1	88.8	98.0	98.3	4.9	11.2	2.0	1.7
Fuel	31.9	28.7	21.6	23.1	68.1	71.3	78.4	76.9
Daily necessities	87.7	96.3	99.5	99.7	12.3	3.7	0.5	0.3

Source: State Statistical Bureau (1986). *China Statistics Yearbook 1986*. Chinese Statistical Publishing House, Beijing; State Statistical Bureau (1988). *Rural Statistical Yearbook of China 1988*. Chinese Statistical Publishing House, Beijing. (In Chinese)

peasants are turning from a semi-self-sufficiency economy to a market economy, although the proportion of food self-supplied is still quite high. As the peasants only have a limited area of cultivated land available per person, their products, including 'grain', are mainly for their own use. It is the surplus that enters circulation as commodities.

The moderately well off and the poor

Most Chinese economists consider that the rise of the people's living standard goes through four stages: hunger, enough to eat and wear, moderately well-off, and well-to-do. China went through the stage of hunger, before Liberation, to the stage of enough to eat and wear, and is now approaching the moderately well-off stage. At present there are no quantitative criteria for the terms poor, or moderately well-off: they are but qualitative descriptions. Poor generally means that people are short of the means to live, and suffering from hunger for an indefinite time. Poverty occurs mainly in the rural areas.

In China the difference between the cities and the countryside is obvious. Before the wide-ranging increases in the prices of agricultural products in 1979, there was a great difference between the incomes of the city residents and of the peasants, due to the unequal exchange-rate between agricultural and industrial products. In addition, the government provided agricultural products to the urban dwellers at low prices, while the purchase price set for buying agricultural products from the peasants was too low. This further increased the difference in the consumption levels of the peasants and urban dwellers (registered non-agricultural population) — from 1:2.4 in 1952 to 1:2.9 in 1978. More recently this difference has been reduced, and in 1985 the ratio was down to 1:2.3. The details are given in Table 6.25, though, in reality, the difference is much greater than that shown in the table.

In Chinese cities, basically, hunger does not exist. Poor peasants are unable to move to the cities, as there are strict controls on the migration of unemployed people. One reason for the lack of poverty in the cities is the implementation of a full employment policy for city youth in order to prevent the development of a poverty stratum. But more importantly, a relatively complete social insurance system is provided by the government for city dwellers. The Staff and Workers (see Appendix 3) in the cities receive low nominal salaries, but they receive a great deal of social welfare in addition, and their employers provide further welfare. This welfare takes the form of public health services, retirement salaries, and subsidised housing, food, fuel, kindergarten and education for children, and so on. In addition,

TABLE 6.25 The consumption levels of the peasants and of the non-agricultural population in 1952, 1978, and 1985

	1952	1978	1985
Whole country (yuan/person)	76	175	407
Peasants (yuan/person)	62	132	324
Non-agricultural population (yuan/person)	148	383	754
Ratio of consumption of peasants to that of non-agricultural population	1:2.4	1:2.9	1:2.3

Source: State Statistical Bureau (1986). *China Statistics Yearbook 1986*. Chinese Statistical Publishing House, Beijing. (In Chinese)

the government and/or business enterprises will give relief if it should be necessary.

At present a considerable proportion of the nation's rural population is on low incomes. There are 4.7×10^6 poor households under state relief or receiving collective allowances, which is 42 per cent of the poor households. According to a report in the *People's Daily*, 23 February 1987, the percentage of the moderately well-off rural households with per caput net incomes of more than 1000 yuan per annum, increased from 2.3 per cent in 1985 to 3.4 per cent in 1986, while the percentage of the poor rural households with per caput net incomes of less than 200 yuan decreased from 12.2 per cent in 1985 to 11.3 per cent in 1986. However, the percentage of poor rural households with per caput net incomes of less than 100 yuan increased slightly from 0.9 per cent in 1985 to 1.1 per cent in 1986.

There are many reasons for the variations in income in the Chinese rural areas. Differences in geographic location, economic situation, and the quality and quantity of cultivated land available in relation to the population, are all very important factors. After the implementation of the household contracted responsibility system, disabled and weak peasants were not able to earn as much as the able-bodied, and, in general, the capabilities of a labourer and the equipment and assets available to him or her now influence the level of income. Higher incomes are also more easily obtained from the secondary and tertiary sectors, where the profits are generally higher than they are from the primary sector.

The differences in income in the Chinese rural areas are most obviously apparent as regional differences. Most of the poverty areas are in the western rural economic region, though some are in the central rural economic region. The great majority of these poverty areas are in the mountains — some are where minority nationalities live, and some are where the bases of the Chinese Communist Party were during the early stages of the revolution.

As shown in Table 6.26, the net income in the eastern rural economic region in 1986 increased at a much faster rate than that in the western rural economic region. The geographical conditions are more favourable in the east, and economic development is more advanced, but even more importantly, the township enterprises there are more developed. Differences in income result in differences in expenditure, and Table 6.27 illustrates the differences in the peasants' per caput consumption of the main food items in five different provinces.

The Chinese government is aware of the poverty in the developing areas and has decided to relax various restrictions further, to reduce rural tax or even provide exemptions from it, and to give financial allowances and technical help in these areas.

TABLE 6.26   The net income per person in the three rural economic regions in 1985, and the increase in net income from 1985 to 1986

Rural economic region	Net income in 1986 (yuan/person)	Increase in net income from 1985 to 1986 (%)
Eastern	495.6	7.1
Central	397.4	2.3
Western	327.9	1.9

Source: *People's Daily*, 23 February 1987. (In Chinese)

TABLE 6.27   The annual consumption of foodstuffs per person in peasant households in Jiangsu, Anhui, Shanxi, Guizhou, and Gansu in 1985

	Jiangsu (kg)	Anhui (kg)	Shanxi (kg)	Guizhou (kg)	Gansu (kg)
'Grain' (raw)	291.2	276.7	223.2	219.1	244.2
Vegetables	117.3	100.6	92.6	151.1	40.5
Vegetable cooking oil	4.2	3.1	2.8	0.6	2.3
Non-vegetable cooking oil	0.5	1.2	0.2	2.8	0.6
Pork, beef and mutton	9.2	8.1	4.3	16.6	6.2
Poultry	1.6	1.8	0.1	0.5	0.1
Eggs	4.2	1.6	2.0	0.7	0.8
Aquatic products	3.0	1.1	0.03	0.2	<0.01
Sugar	1.8	1.5	1.0	0.9	0.3
Alcoholic drinks	5.9	2.9	1.2	4.8	1.0

Source: State Statistical Bureau (1987). *Rural Statistical Yearbook of China 1986*. Chinese Statistical Publishing House, Beijing, (In Chinese); Editorial Board of China Agriculture Yearbook (1987). *China Agriculture Yearbook 1986 (English Edition)*. Agricultural Publishing House, Beijing.

References

The following texts are in Chinese:

Editorial Board of China Agriculture Yearbook (1987). *China Agriculture Yearbook 1986*. Agricultural Publishing House, Beijing.

Department of Rural Statistics, State Statistical Bureau (1986). *Rural Statistical Yearbook of China 1986*. Chinese Statistical Publishing House, Beijing.

Industry and Enterprise Division, Development Institute (1986). On the development of non-agricultural sectors in the rural areas. *Economic Research*, nos 8 and 9.

Policy Studies Division, Ministry of Agriculture, Animal Husbandry, and Fishery (1982). *A Sketch of the Chinese Rural Economy*. Agricultural Publishing House, Beijing.

State Statistical Bureau (1986). *China Statistics Yearbook 1986*. Chinese Statistical Publishing House, Beijing.

State Statistical Bureau (1989). *Selected Superior Papers on Statistical Analysis*. Chinese Statistical Publishing House, Beijing.

Year 2000 Research Group (1984). *China in the Year 2000*. Science and Technology Literature Publishing House, Beijing.

The following texts are in English:

Anon. (1984). *Life in Modern China*. China Spotlight Series. New World Press, Beijing.

Barker, Randolph and Sinha, Radha with Rose, Beth (Eds) (1982). *The Chinese Agricultural Economy*. West Special Studies in China and East Asia. Westview Press, Boulder CO.

Editorial Board of China Agriculture Yearbook (1986). *China Agriculture Yearbook 1985 (English Edition)*. Agricultural Publishing House, Beijing.

Editorial Board of China Agriculture Yearbook (1987). *China Agriculture Yearbook 1986 (English Edition)*. Agricultural Publishing House, Beijing.

World Bank (1986). *World Development Report 1986*. Oxford University Press, New York.

7

Agricultural education, research, and extension

LI XIAOCHUN

Agricultural education, research, and extension are carried out in three separate systems, though all are under the control of the Ministry of Agriculture, Animal Husbandry, and Fishery (now renamed the Ministry of Agriculture), and quite a number of scientists are engaged in two, and sometimes three of these activities.

Agricultural education

Higher agricultural education

In China the history of higher education in agriculture is quite short, and extends back only to the last decade of the nineteenth century. The first formal agricultural school in China was established in Wuchang, Hubei province, in 1898. Other agricultural schools were then set up in Baoding in 1902, Beijing and Chengdu in 1905, Jinan in 1906, Hangzhou in 1910, and Nanjing in 1913. These schools were the pioneers of agricultural education in China. All have experienced many changes since their establishment, so it is best to say simply that they were the predecessors, respectively, of the Central China Agricultural University, Hebei Agricultural University, Beijing Agricultural University, Sichuan Agricultural University, Shandong Agricultural University, Zhejiang Agricultural University, and Nanjing Agricultural University.

1952 was of great importance in Chinese education, for in that year there was a nationwide readjustment in all universities and colleges, including agricultural colleges. Before then most of the agricultural colleges had been part of a university, but after the readjustment they became independent institutions, and Beijing Agricultural University was established as the first agricultural university. After 1952 many new agricultural institutes were established, and many of these became agricultural universities. It should, perhaps, be added that in China an agricultural institute and an agricultural university have the same level of students, lectures, and duration of study, and students taking higher level courses will all graduate at bachelor level.

They are, in fact practically the same except in name. The term institute is frequently used in the name of a teaching institution, and also in the name of an institution that only undertakes research. The teaching institutes and the schools (see below) are also often referred to, collectively, as colleges, and the word college still features in the English version of the name of some of the agricultural higher education institutions.

There were not many agricultural colleges in China from the beginning of the 1920s to the end of the 1940s, but some of those that did exist were quite famous. The best known was the Agricultural College of the University of Nanking, especially its Departments of Agricultural Economics and Pest Control.[1] The next most famous agricultural college was that of the University of Zhongyang (which means central), where the Department of Animal Husbandry and Veterinary Science, and the Department of Agronomy, were especially renowned. After the readjustment of higher education these two agricultural colleges were combined to form the Nanjing Agricultural Institute, which is now the Nanjing Agricultural University. The Departments of Sericulture and of Pest Control at the Agricultural College of Zhejiang University were also well known, and they now form part of Zhejiang Agricultural University. In addition, the Agricultural College of Zhongshan University in Guangzhou (now the South China Agricultural University); the Agricultural College of Sichuan University in Chengdu (now the Sichuan Agricultural University in Yaan, Sichuan province); and the North-West Agricultural College in Wugong, Shaanxi province (now the North-West Agricultural University), were all considered to have an important influence in their local regions.

There were 69 agricultural institutions of higher education in China under the control of the Ministry of Agriculture, Animal Husbandry, and Fishery in 1985 (the Ministry of Forestry also controlled equivalent institutions for higher education in forestry). Of these, 47 were universities and institutes of agriculture, aquatic products, or agriculture and land reclamation; the remainder were agricultural schools which taught to a lower level than the universities and institutes. The 47 agricultural universities and institutes were distributed throughout China, with at least one such institution in each province, autonomous region, or municipality (Table 7.1).

Table 7.1, lists eight key agricultural universities. One is the Beijing Agricultural Engineering University, which is the only agricultural university

1. The University of Nanking disappeared after the readjustment of higher education; and the present University of Nanjing was established after that readjustment. Nanking was the English translation under the old system for the name of the city that is now translated as Nanjing; before 1949 this city was the national capital. The Chinese name of the old University of Nanking was 'Jinling Daxue' wherein Jinling was the name of the ancient city of Nanjing and Daxue meant university. The present University of Nanjing uses just the present name of the city.

TABLE 7.1   The agricultural institutions of higher learning in China in 1985

Name	Location	Control	Enrollment
Beijing Agricultural University*	Beijing	MAAHF	1639
Beijing Agricultural Engineering University*	Beijing	MAAHF	1433
Beijing Agricultural Institute	Beijing	MAAHF	670
Tianjin Agricultural Institute	Tianjin	Tianjin	371
Hebei Agricultural University	Baoding, Hebei	Hebei	3001
Zhangjiakou Agricultural School	Xuanhua, Hebei	Hebei	640
Hebei Agricultural Teachers College	Changli, Hebei	Hebei	332
Shanxi Agricultural University	Taigu, Shanxi	Shanxi	2726
Shanxi Linfen Agricultural School	Linfen, Shanxi	Shanxi	303
Inner Mongolia Institute of Agriculture and Animal Husbandry	Hohhot, Inner Mongolia	Inner Mongolia	3353
Zhelimu Institute of Agriculture and Animal Husbandry	Tongliao, Inner Mongolia	Inner Mongolia	667
Shenyang Agricultural University*	Shenyang, Liaoning	MAAHF	2469
Shenyang Agricultural Institute	Shenyang, Liaoning	Liaoning	307
Xionyue Agricultural School	Gaixian, Liaoning	Liaoning	384
Dalian Institute of Aquatic Products	Dalian, Liaoning	MAAHF	1053

TABLE 7.1   (*Cont.*)

Name	Location	Control	Enrollment
Jilin Agricultural University	Changchun, Jilin	Jilin	2929
Yanbian Agricultural Institute	Longjing, Jilin	Jilin	1187
Jilin Special Products School	Jilin, Jilin	MAAHF	80
North-East Agricultural Institute	Harbin, Heilongjiang	Heilongjiang	1513
Heilongjiang August 1st University of Agriculture and Reclamation	Mishan, Helongjiang	Helongjiang	1440
Heilongjiang Teachers School of Agriculture and Reclamation	Acheng, Heilongjiang	Heilongjiang	989
Shanghai Agricultural Institute	Shanghai	Shanghai	893
Shanghai University of Aquatic Products	Shanghai	MAAHF	1546
Nanjing Agricultural University*	Nanjing, Jiangsu	MAAHF	2236
Jiangsu Agricultural Institute	Yangzhou, Jiangsu	Jiangsu	2314
Suzhou Sericulture School	Wuxian, Jiangsu	Jiangsu	748
Nanjing Agriculture School	Nanjing, Jiangsu	Jiangsu	281
Zhejiang Agricultural University	Hangzhou, Zhejiang	Zhejiang	3245
Hangzhou Agriculture School of Zhejiang Agricultural University	Hangzhou, Zhejiang	Zhejiang	242
Zhejiang Institute of Aquatic Products	Pudu, Zhejiang	MAAHF	802

Name	Location	Control	Enrollment
Zhejiang Teachers School of Rural Technique	Jinxian, Zhejiang	Zhejiang	510
Anhui Agricultural University	Hefei, Anhui	Anhui	2605
Hefei Institute of Agricultural Economics	Hefei, Anhui	Anhui	1003
Fujian Agricultural Institute	Fuzhou, Fujian	Fujian	2423
Xiamen Institute of Aquatic Products	Xiamen, Fujian	MAAHF	752
Jiangxi Agricultural University	Nanchang, Jiangxi	Jiangxi	2746
Yichun Agricultural School	Yichun, Jiangxi	Jiangxi	161
Shandong Agricultural University	Tai'an, Shandong	Shandong	3607
Laiyang Agricultural Institute	Laiyang, Shandong	Shandong	1702
Henan Agricultural University	Zhengzhou, Henan	Henan	2417
Baiquan Agricultural School	Huixian, Henan	Henan	1187
Yuxi Agricultural School	Xin'an, Henan	Henan	728
Yunan Agricultural School	Xinyang, Henan	Henan	962
Zhengzhou School of Animal Husbandry and Veterinary Medicine	Zhengzhou, Henan	Henan	1115
Central China Agricultural University*	Wuhan, Hubei	MAAHF	2788
Hunan Agricultural Institute	Changsha, Hunan	Hunan	2551

TABLE 7.1   (*Cont.*)

Name	Location	Control	Enrollment
South China Agricultural University*	Guangzhou, Guangdong	MAAHF	3201
Zhanjiang Agricultural School	Zhanjiang, Guangdong	Guangdong	915
Zhongkai Agricultural Technology Institute	Guangzhou, Guangdong	Guangdong	83
Foshan Veterinary School	Nahai, Guangdong	Guangdong	779
Zhanjiang Institute of Aquatic Products	Zhanjiang, Guangdong	MAAHF	875
South China Institute of Tropical Plants	Danxian, Guangdong	Guangdong	1221
Guangxi Agricultural Institute	Nanning, Guangxi	Guangxi	2255
South-West Agricultural University*	Chongqing, Sichuan	MAAHF	3010
Sichuan Agricultural University	Ya'an, Sichuan	Sichuan	1966
Mianyang Agricultural School	Mianyang, Sichuan	Sichuan	1045
Xichang Agricultural School	Xichang, Sichuan	Sichuan	1070
Sichuan Institute of Animal Husbandry and Veterinary Medicine	Rongchang, Sichuan	Sichuan	725
Guizhou Agricultural College	Guiyang, Guizhou	Guizhou	2129
Yunnan Agricultural University	Kunming, Yunnan	Yunnan	1928
Tibet Institute of Agriculture and Animal			

Name	Location	Control	Enrollment
Husbandry	Linlan, Tibet	Tibet	189
North-West Agricultural University*	Wugong, Shaanxi	MAAHF	2828
Gansu Agricultural University	Lanzhou, Gansu	Gansu	1835
Qinghai Institute of Animal Husbandry and Veterinary Medicine	Xining, Qinghai	Qinghai	494
Ningxia Agricultural Institute	Yinchuan, Ningxia	Ningxia	1284
Xinjiang August 1st Agricultural Institute	Ürümqi, Xinjiang	Xinjiang	3178
Shihezi Medical College	Shihezi, Xinjiang	MAAHF	1023
Tailimu University of Agriculture and Reclamation	Aksu, Xinjiang	MAAHF	1033
Shihezi Agricultural Institute	Shihezi, Xinjiang	MAAHF	1733

* A key institution.

MAAHF = Ministry of Agriculture, Animal Husbandry and Fishery (now called Ministry of Agriculture).

Source: Editorial Board of China Agriculture Yearbook (1987). *China Agriculture Yearbook 1986*. Agricultural Publishing House, Beijing. (In Chinese)

that specializes in engineering, and the other seven are related to a particular 'big region'. A 'big region' was an administrative division which existed in the early 1950s and which was placed at a level between the central administration and the provinces. Although these divisions have long since been abandoned, their influence can still be found here and there. In this instance, the Shenyang Agricultural University is within the boundary of the North-East China 'big region'; the Beijing Agricultural University is in the North China 'big region'; the Nanjing Agricultural University is in the East China 'big region'; the Central China Agricultural University is in the Central China 'big region'; the South China Agricultural University is in the South China 'big region'; the South-West Agricultural University is in the

South-West China 'big region'; and the North-West Agricultural University is in the North-West China 'big region'. The Ministry of Agriculture, Animal Husbandry, and Fishery separated out these key universities for preferment so that the Ministry officers could concentrate their efforts on improving them and so set standards for other institutions to emulate.

Degree level students in the universities and institutes study for four years (five years for veterinary science) and receive a bachelor's degree on graduation. Many of these institutions also offer postgraduate training; in addition, many offer sub-degree level training similar to that offered by the higher education schools. These agricultural schools, including the schools of agriculture and land reclaimation, offer shorter courses of two or three years' duration, which are at a lower level than the degree courses in the universities and institutes, and for which the graduates do not receive degrees. They tend to offer a narrower range of courses, and on more specialized subjects.

The students who enter any of these institutions of higher education are enrolled after taking the nationwide admission examination. The students are separated into different departments at the beginning of their studies and they follow their department's courses thereafter. The courses which the students are to take are decided upon by the staff of the institutions concerned and are then approved by the Ministry. In most cases there is very little scope for students to make selections between courses — there are now moves towards allowing students some choice, but these are still at an experimental stage and are not yet popular.

Generally speaking, the students study basic courses such as mathematics, physics, and a foreign language in their first and second years, basic technical courses in their second and third years, and take courses related to their specialty in their third and fourth years. All students have to learn a foreign language which, in most cases at present, is English. They also have to undertake a group project, and possibly write a short thesis, before they complete their course. On graduation all students are assigned by the government to work in a government organization, university, institute, school, or research institution.

Many of the faculty members are engaged in research as well as teaching. The current policy of the government is that all faculty members should treat teaching as their main task, but should also undertake research. Recently it has been emphasized that faculty members should become involved in extension work as well, and many higher education institutions have been trying to make contact with local agricultural production units — the aim is to establish a three-in-one basis of teaching, research, and extension.

Since the end of the 1970s the funds provided by the government for higher education in agriculture have been increased substantially, and some universities and institutes have also received loans from the World Bank. This has resulted in a great increase in the provision of equipment and the training of faculty members, and a number of academics have also been able to go abroad to visit or study. Links with overseas institutions have been encouraged and many individual universities and institutes have now established mutual relations with comparable bodies overseas. Some exchange scholars for visits and to give lectures, others undertake joint research projects or provide assistance in establishing new specialities, while a recent venture is the joint training of postgraduate students. All these developments have taken place during the 1980s.

Education in China, including agricultural education, is under the leadership of the Education Committee of the State Council, which is responsible for educational undertakings throughout China. Within this overall administrative structure, agricultural education is directly administered by the Education Department of the Ministry of Agriculture, Animal Husbandry, and Fishery. The corresponding organizations in the provinces, autonomous regions, and municipalities are responsible for agricultural education in their areas. In Table 7.1 it will be seen that some of the agricultural universities or institutes are under the direct control of the Ministry of Agriculture, Animal Husbandry, and Fishery but most are responsible to the corresponding provincial level organization.

Of all the institutions of higher education in agriculture in China, the Beijing Agricultural University has received the most attention from the Ministry of Agriculture, Animal Husbandry, and Fishery. It has the most able faculty members, and far more associate professors and professors than in other agricultural universities. Similarly, the financial support, and the opportunities for staff contact with overseas colleagues, are far superior. It is a multi-disciplinary university and holds a leading position in the country in the fields of biotechnology, animal husbandry, veterinary science, and agricultural economics.

Nanjing Agricultural University was established on the basis of the once famous Agricultural College of the University of Nanking, and of the Agricultural College of Zhongyang University. At present it is quite strong in plant protection, veterinary science, and southern soya bean research.

The South China Agricultural University was established on the basis of the Agricultural College of Zhongshan University and the Agricultural College of Lingnan University. It holds a leading position in the fields of tropical crop cultivation, agricultural ecology, and sericulture. It also has the largest campus in China — covering about 500 ha.

The origins of the Central China Agricultural University go back to the Wuchang Agricultural School, so it has the longest history of all the higher education agricultural institutions in China. Its campus covers more than 200 ha and includes hills, lakes, woods, and grasslands in a beautiful landscape. The university is particularly strong in the fields of food science, edible fungi, and agricultural micro-organisms.

As before, the North-West Agricultural University still holds the leading position in farming under arid and semi-arid conditions.

Zhejiang Agricultural University ranks first in the application of radio-isotope techniques in agriculture in China, and it has also for long had a high reputation in sericulture.

The North-East Agricultural University enjoys long standing leadership in northern soya bean research, and it is also well known for its work in agricultural mechanization and animal husbandry. The campus, which was rebuilt after being destroyed, is also recognized as one of the best in the country.

Other universities and the fields in which they have a particular reputation are: the South-West Agricultural University, for tea research; the Gansu Agricultural University, for grassland ecology; the Inner Mongolia Institute of Agriculture and Animal Husbandry, for grassland ecology; the Sichuan Agricultural University, for poultry husbandry; and the Hunan Agricultural Institute, for bast-fibre and tea research. The Fujian Agricultural Institute is not only strong in apiculture and tropical fruit studies, but also has a beautiful, newly-built campus, which is one of the best in the country.

In addition, it should be mentioned that the Shanghai University of Aquatic Products is the leading institution of the six concerned with aquatic products; and the Shihezi Agricultural Institute is the leading institution of the six concerned particularly with agriculture and land reclamation. And, of course, there is the Beijing Agricultural Engineering University, which is the only university specializing in agricultural engineering, and which is one of the key universities.

Secondary schools specializing in agriculture

There are 470 secondary schools specializing in agriculture, at least one for each prefecture. The total enrolment in 1985 was 136 000. These schools offer specialized courses in agronomy, animal husbandry, veterinary practice, agricultural land reclamation, forestry, water conservation, aquaculture, agricultural mechanization, horticulture, sericulture, tea production, agricultural economics, meteorology, and other similar subject areas. They enrol students who have completed middle school and the students study for a further three years. Usually they enrol local students, and the students are assigned to work locally after the completion of their courses — most of the cadres working in agriculture at county level have been trained at these specialized schools. Generally, the schools specializing in agriculture engage in some field experimental work and agricultural extension.

There are also some middle schools which include agriculture in their curriculum. Some include specific lessons in agriculture, others just teach some agricultural knowledge related to the agricultural activities in their locality. These school are mainly in the countryside, and their intention is to give middle school students some agricultural knowledge in order to help them with their agricultural activities after they leave school. Some arrange for a considerable amount of practical work to be done and practice a system of half practical work and half classroom teaching. In 1985 there were 6555 of these so-called agricultural middle schools, or agricultural specialized or professional schools in China.

Adult agricultural education

In China there are 29 provinces, autonomous regions, and municipalities; hundreds of prefectures and cities; more than 1900 counties; and over 30 000 townships. Each different administrative level has its agricultural organizations, so the number of cadres engaged in agricultural administration is very large. For historical reasons many of them have not received enough agricultural education. In order to raise the level of the cadres' agricultural knowledge, and/or to bring them up to date, the Ministry of Agriculture, Animal Husbandry, and Fishery established the Cadres' Institute for Agricultural Management, and arranged for some agricultural higher education institutions to set up Branch Institutes of the Cadres' Institute for Agricultural Management to train cadres involved in agricultural management. It is reported that between 1981 and 1985 over 20 000 cadres in charge of agriculture, who were employed in administrative units at a level above county level, received training. In addition to the Cadres' Institute there are other cadres' schools and training courses set up at the provincial and prefecture levels which train cadres locally.

Agricultural training is also provided for the peasants. As the economic and living standards of the peasants have improved, their interest in technical training has also increased. Numerous short training courses have been arranged for the peasants, and they have also attended courses at junior and secondary technical schools. The various courses have offered specialized technical training in poultry raising, aquaculture, cotton cultivation, and many similar topics. The courses are short, and aim to provide broad mass education. In 1985 just under 5 million people received junior or secondary technical education and some 31 million attended short-term courses. Literacy training is also provided for the peasants: at the time of the 1982 census 37 per cent of the people working in agriculture were illiterate.

The Central Agricultural Broadcasting School was established in 1980 to provide another form of adult education, and is sponsored by the Ministry of Agriculture, Animal Husbandry, and Fishery. Teachers are invited from various institutions, according to their fields of expertise, to produce radio and television courses, and at the same time auxiliary teaching corps are organized in different places all over the country to help people with their learning. Forty courses in the fields of agronomy, agricultural management, animal husbandry, and freshwater fish raising are offered. Teaching is arranged over three years, and those who are successful in the examinations receive a diploma which is the equivalent of the awards received by the graduates of the secondary agricultural technical schools. In 1985 there were 830 000 students enrolled in the Central Agricultural Broadcasting School, and 70 000 graduated.

Agricultural research

Agricultural research in China is mainly the province of a special system of research organizations, though important work is also carried out in the universities and teaching institutes as mentioned above. The institutes of the Chinese Academy of Sciences, the Ministry of Forestry, and the Ministry of Water Resources and Electric Power also undertake some research relevant to agriculture. In addition, these various groups of scientists undertake joint projects as appropriate.

The first agricultural research institution established in China was the Agricultural Experiment Station of Beijing, founded in 1906. The main institutions in existence at the time of Liberation in 1949 were the Central Agricultural Experiment Institute (founded in 1931), the Animal Husbandry Experiment Institute, and the Forestry Experiment Institute, all of which were in Nanjing, the capital at that time; and several agricultural experimental stations situated in North-East China, North China, and South-West China. In the early 1950s the former Ministry of Agriculture formed seven agricultural research institutes from the agricultural research institutions which already existed in the seven 'big region' administrative units. Then in 1957 the Chinese Academy of Agricultural Sciences was founded in Beijing. By 1985, in addition to the Head Office and Library, the Academy had 32 research institutes and divisions employing 10 504 staff. Of these institutes and divisions, 13 are in Beijing and the other 19 are in localities of particular relevance to their activities. The Institute of Cotton Science, for instance, is in Anyang, Henan province, the centre of a major cotton producing region; the Institute of Sugar Beet is in Hulan, Heilongjiang province, the centre of a major sugar beet region; and the Institute of Tobacco Science is in Yidu, Shandong province, where Qingzhou tobacco comes from. The Department of Agricultural History, Nanjing Agricultural University, is also associated with the Chinese Academy of Agricultural Sciences. Details of all the institutions which are part of the Chinese Academy of Agricultural Sciences are given in Table 7.2.

In addition to the above, all the provinces, autonomous regions, and municipalities have their own provincial academy of agricultural sciences. Seven of these were developed from the Agricultural Research Institutes of the former 'big regions' of the early 1950s, and the remainder have been established successively over the last 30 years. The research work of these academies is under the direction of the Chinese Academy of Agricultural

TABLE 7.2   The institutes and laboratories of the Chinese Academy of Agricultural Sciences in 1985

Name	Location	Scientific staff
Institute of Soils and Fertilizers	Beijing	223
Institute of Crop Breeding and Cultivation	Beijing	228
Institute of Crop Germplasm Resources	Beijing	173
Institute of Plant Protection	Beijing	163
Institute for the Application of Atomic Energy in Agriculture	Beijing	206
Institute of Animal Science	Beijing	178
Institute of Vegetable Crops	Beijing	141
Institute of Agricultural Economics	Beijing	131
Institute of Agricultural Natural Survey and Regional Research	Beijing	70
Institute of Information on Agricultural Science and Technology	Beijing	182
Institute of Apiculture	Beijing	94
Laboratory of Agricultural Meteorology	Beijing	64
Laboratory of Biological Control	Beijing	50
Library	Beijing	37
Institute of Cotton	Anyang, Henan	164
Institute of Oil Crops	Wuchang, Hubei	152
Xingcheng Institute of Pomology	Xingcheng, Liaoning	129
Zhengzhou Institute of Pomology	Zhengzhou, Henan	153
Institute of Citrus	Chongqing, Sichuan	101
Institute of Sugar Beet	Hulan, Heilongjiang	79
Institute of Tea	Hangzhou, Zhejiang	108
Institute of Tobacco	Yidu, Shandong	85
Institute of Jute and Kenaf	Yuanjiang, Hunan	70
Institute of Irrigation and Drainage	Xinxiang, Henan	137
Lanzhou Institute of Animal Science	Lanzhou, Gansu	93
Institute of Grasslands	Hohhot, Inner Mongolia	148
Harbin Veterinary Research Institute	Harbin, Heilongjiang	202
Lanzhou Veterinary Research Institute	Lanzhou, Gansu	177
Lanzhou Institute of Traditional Chinese Veterinary Medicine	Lanzhou, Gansu	141
Southern Institute of Animal Parasitology	Shanghai	64
Institute of Sericulture	Zhenjiang, Jiangsu	154
Institute of Special Plant and Wildlife Utilization	Yongji, Jilin	243
Chinese National Rice Research Institute	Hangzhou, Zhejiang	133

Source: From a general survey held by the Ministry of Agriculture, Animal Husbandry, and Fishery in 1986.

Sciences, but the administration is under the control of the local provincial government. These provincial academies are comprehensive in coverage. In addition to them, there are many specialized research institutions established to meet the particular requirements of a province in more limited fields such as cash crop production, agricultural mechanization, and so on. In 1985 there were 428 provincial, autonomous region, and municipal research institutions for agriculture, animal husbandry, and fishery, employing 63 594 staff, of whom 46 552 were scientific and technical staff.

For historical reasons there are quite large differences between the provincial academies of agricultural sciences in their capabilities to undertake agricultural research. Generally speaking, the Jiangsu Academy of Agricultural Sciences in Nanjing has the highest academic level because it was established from what was the former Central Agricultural Experimental Institute and other related organizations before 1949; these had become the East China Agricultural Research Institute in the 1950s, and the Jiangsu Branch of the Chinese Academy of Agricultural Sciences in the 1970s. After the Jiangsu Academy, most people believe, comes the Jilin Academy of Agricultural Sciences. Its predecessor was the Experimental Station of the Manchurian Railroad Industry, established in 1913, which afterwards became the Gongzhuling Agricultural Station, and then the Agricultural Research Institute of the North-East 'big region' in the early 1950s.

There are also agricultural experimental institutions at prefecture level, and in 1985 there were 658 of these institutions employing 32 600 technical staff.

In addition to the Chinese Academy of Agricultural Sciences, there are also the following research institutions at national level under the control of the Ministry of Agriculture, Animal Husbandry, and Fishery: the Chinese Academy of Aquatic Products, the South China Academy of Tropical Plants, the Chengdu Institute of Marsh Gas Science, the Nanjing Institute of Agricultural Mechanization, and the Environmental Protection and Monitoring Institute.

The headquarters of the Chinese Academy of Aquatic Products is situated in Beijing. Under the Academy are three Aquatic Products Institutes for three river valleys — the Heilong Jiang River Valley, the Yangtze River Valley, and the Pearl River Valley; three Aquatic Products Institutes for three seas — the Yellow Sea, the East China Sea, and the South China Sea; and three special institutes — the Institute for Fishing Instruments, the Institute for Fishing Engineering, and the Institute for Freshwater Fishing. In 1985 3 480 people were employed in the Academy and its institutes, of whom 1 679 were scientific staff.

The South China Academy of Tropical Plants is a research centre for tropical plant cultivation, with its main emphasis on rubber. Its head office is located in Danxian, Hainan Dao Island, Guangdong province. The

Academy is responsible for six research institutes which undertake investigations into rubber plant cultivation and other issues, and for three experimental stations situated in Xinglong, Yuexi, and Wenchang. The station in Wenchang deals with coconut production. In 1985 the Academy and its institutions employed 2 233 people of whom 643 were scientific staff.

The Chengdu Institute of Marsh Gas Science is in Chengdu, Sichuan province, and it had 102 scientific staff in 1985; the Environmental Protection and Monitoring Institute is in Tianjing, and it had 83 scientific staff in 1985; and the Nanjing Institute of Agricultural Mechanization is in Nanjing, and it had 272 scientific staff in 1985.

Finally, there is the Chinese Academy for Agricultural Engineering Research and Design which is also under the control of the Ministry of Agriculture, Animal Husbandry, and Fishery. It is in Beijing and is mainly concerned with design work, though it undertakes some research. In 1985 it had 267 technical staff.

Since 1983 the Ministry of Agriculture, Animal Husbandry, and Fishery has created nine new agricultural research institutes, which work at an advanced technological level. These are the Central Laboratory of the Sichuan Academy of Agricultural Sciences, the Agricultural Analysis and Measurement Centre of the Hubei Academy of Agricultural Sciences, the Agricultural Genetics and Physiology Institute of the Jiangsu Academy of Agricultural Sciences, the Institute of Rice Science of the Guangdong Academy of Agricultural Sciences, the Institute of Soya Bean Science of Jilin Academy of Agricultural Sciences, the Central Laboratory of the Xinjiang Academy of Agricultural Sciences, the Mizhi Experimental Station of the Loess Plateau Integrated Control of the Shaanxi Academy of Agricultural Sciences, the Central Analytical Laboratory of the Institute of Soils and Fertilizers of the Chinese Academy of Agricultural Sciences, and the National Germ-Plasm Bank of the Chinese Academy of Agricultural Sciences.

The Ministry of Agriculture, Animal Husbandry, and Fishery has also invested jointly with various local governments to reinforce selected research institutions, such as the Institute of Millet Research in Hebei province, and the Institute of Sweet Potato Research in Xuzhou, Jiangsu province. In addition, the State Planning Committee has given financial support to the Hunan Academy of Agricultural Sciences, which has had considerable success in rice hybridization, to build a modernized Rice Hybridization Institute. Improvements of this nature are still under way.

The institutes of the Chinese Academy of Sciences which are related to agriculture are the Institute of Genetics, the Institute of Zoology, the Institute of Microbiology, the Institute of Pedology, the Institute of Soil and Water Conservation, and the Institute of Geography. There are also some other research institutes under other ministries, including the Ministry of Forestry, the Ministry of Water Resources and Electric Power, and the

Machinery Committee, which have some projects related to agriculture, particularly concerning soil and water conservation, irrigation and drainage, agricultural machinery, and fruit and flower growing.

The roles of the various research institutions vary according to their level in the administrative hierarchy. The research institutes directly subordinate to the ministries, together with the universities and teaching institutes, engage in basic and applied research, with due regard to the potential for practical application. The provincial research institutions undertake applied research and local trials, and if conditions are appropriate may also do some more-basic research. The experimental organizations at prefecture level are mainly responsible for the introduction and application of new techniques which will be useful locally. They may conduct some field trials if conditions are suitable. Communication and cooperation between the different research and experimental institutions is, of course, very important.

The funds and programmes of the research institutions at all levels are arranged by their higher administrative authorities; sometimes these authorities request that specific research projects are undertaken, for which they will then provide additional funds. In 1985 the Central Committee of the Communist Party of China promulgated the Decision about the Reform of the Scientific and Technical System. This asked all the research institutions to ally their research more closely with production, and in this way to serve the development of production and of the commodity economy more effectively. As a result reforms have been undertaken in the planning and management of scientific research, and emphasis has recently been given to easing restrictions and enlivening the research institutions.

In addition to this general funding and direction from the administrative authorities, there are various other sources of funds and projects for the research institutions. Institutions of high academic standing wishing to pursue basic studies, or applied research at a fairly basic level, can apply to the State Fund for Natural Science for funds for particular research projects. Institutions wishing to undertake applied research projects and local trials can apply to the relevant department to have a project or trial designated as a national key project, which would make it a specially promoted project of that department. If, after discussions and examination, the application is accepted, then the project will be assigned formally to the institution which had made the application and funds will be provided to carry it out. At present research institutions are being particularly encouraged to strengthen their relations with production units and enterprises with a view to their providing technical consultations and services, and also being entrusted with research projects by these units and enterprises, which would pay for them. In this way the research organizations will be able to earn additional funds, and this source of funds is gradually expanding.

There are many rewards and prizes for achievement in scientific research in China. The important ones are the State invention prize, the natural science prize, the State prize for progress in science and technology, and the prizes for progress in science and technology awarded by the ministries and other official bodies, and by the provinces. There are four grades for the State invention prize and natural science prize. Besides the certificate of merit, the scientist awarded the prize will also receive 20 000, 10 000, 5000, or 2000 yuan according to the grade of his prize. If an achievement is exceptional then it can be rewarded with a special prize which will not be limited by the normal standards. The national prizes for progress in science and technology are divided into three grades and a scientist awarded one of these prizes will receive a certificate as well as 15 000, 10 000, or 5000 yuan, according to the grade of his prize. The prizes for progress in science and technology awarded by the ministries and provinces are also divided into three grades and a scientist receiving one of these prizes will receive a certificate and 5000, 3000 or 1500 yuan, according to the grade of his prize. All these prizes for achievement in science and technology are determined by set procedures. Firstly achievements are reported to the State Committee for Science and Technology, or to the relevant department, by the responsible authority. Then the achievements are examined by a specific evaluation committee appointed by the State Committee for Science and Technology or by the relevant department. The results of these examinations are checked by the State Committee or relevant department, and finally the prizes are awarded, usually once a year.

Agricultural extension

Agricultural extension in China is the responsibility of a specific agricultural extension system, but the agricultural research system and the agricultural education system also contribute to extension.

There is a General Extension Station of Agricultural Technology in the Ministry of Agriculture, Animal Husbandry, and Fishery, which oversees extension work throughout the country; in the provinces and prefectures there are corresponding organizations which are responsible for agricultural extension within these administrative areas. Only the county level extension organizations and their agencies — the agricultural extension stations in the districts — directly carry out extension work.

In the early 1970s a four-level agricultural extension system was formed, with the four levels being the county, the commune, the production brigade, and the production team. At that time there was an agricultural experimental institute in each county; an agricultural experimental station in every commune; an agricultural experimental group in each production brigade; and seed plots, experimental plots, and high yielding plots were established by the production teams. In addition various stations had been established at county level, and these included soil and fertilizer stations, plant protection stations, animal breeding stations, and veterinary stations, and also technical training schools. These institutions all provided technical guidance for the peasants.

From the beginning of the 1980s this situation was changed as part of the rural economic reforms. The communes, brigades and teams ceased to exist, and with them went the old system of agricultural extension. All the former county level institutions listed above were brought together in a unified agricultural centre for science and technology with responsibility for experimental work, demonstrations, extension, and training in agriculture throughout the county.

Below county level, there is an agricultural extension station in each district which is an agency of the agricultural centre for science and technology of the county. The main task of this district extension station is to guide and help the agricultural service stations, at township level, to provide extension, technical services, and contract services. The township service stations are managed collectively by a group of selected peasants. At village level, agricultural extension is taken care of by one or more (usually only one) peasant technicians who are selected from the peasants of that village. A peasant technician has the right to receive the same allowance as a village cadre in return for doing his extension work.

A peasant technician can organize specialized households of science and technology to form technical service groups which are then able to provide technical services and undertake technical contracts. For example a group may own a tractor and could provide services where a tractor is needed, or the group may contract to take care of all pest problems in a specific field of crop, or to apply fertilizer to a particular area or crop. All these services have to be paid for by the peasant receiving the service. Agricultural extension in China at the peasant level is basically a transfer of technology — through the provision of contracted and other technical services, consultation, and training — by the more technically advanced peasants to their neighbours in return for payment.

To provide extension services in animal husbandry and veterinary matters, unified animal husbandry and veterinary service centres are gradually being formed at county level by combining the original veterinary stations, animal breeding stations, grassland stations, and other similar institutions. In each county there is also an agricultural machinery service station which is responsible for the various agricultural machinery services provided in the county. Irrigation techniques are taken care of by the Ministry of Water Resources and Electric Power.

The specific agricultural techniques and materials that peasants are made aware of through the extension system are primarily determined by the administrative authorities. These techniques will have arisen from successful research carried out in the research and/or higher education systems. Two examples are the extension of hybrid rice in southern China, and of new varieties of cotton in North China. Sometimes new technology is introduced from abroad; at other times new methods may have to be developed as a consequence of introducing a new fertilizer or pesticide.

When scientists in research institutions, and teachers in higher education institutions and agricultural schools, have obtained useful results from their investigations, they will often carry out trials to test them further. These activities may be considered as a form of extension as they illustrate the results and methods of operation to all who have participated in the work and to peasants living in the area. Sometimes it is necessary to conduct these trials in several places and over a number of years, and in these circumstances the impact will be even greater.

When successful research results have been obtained, the research institution concerned will report the results to the agricultural authority. The technical committee of the Ministry of Agriculture, Animal Husbandry, and Fishery examines these reports regularly. If the results are very good the person who did the work will be given a prize, and the new technology will be passed on to the extension system for wide dissemination if it is appropriate for extension. In the provinces similar technical committees follow similar procedures. Irrespective of whether or not the research earns a prize, the researcher will write papers and articles about the work for publication in journals and bulletins so that the technology related to the research can be selected if needed in practice.

At the same time officers of agricultural administrative units at all administrative levels will consult the relevant specialists if some technical problem arises in their locality. They will then decide on the appropriate technology to deal with the problem in the light of the advice they have received and according to the local conditions, and will pass it on to the local agricultural extension organizations for extension to the peasants.

References

The following texts are in Chinese:

Editorial Board of A Brief Introduction to the Institutions of Higher Education in China (1982). *A Brief Introduction to the Institutions of Higher Education in China*. Educational Science Press, Beijing.

Bureau of Science and Technology, Ministry of Agriculture, Animal Husbandry, and Fishery (1982). *A Sketch of the Agricultural Research Institutions in the People's Republic of China in 1982*. Agricultural Publishing House, Beijing.

Editorial Board of China Agriculture Yearbook (1987). *China Agriculture Yearbook 1986*. Agricultural Publishing House, Beijing.

Information Institute, Chinese Academy of Agricultural Sciences (1984). *An Outline of the Research Institutions of Agriculture and Animal Husbandry at all Levels in China*. Chinese Academy of Agricultural Sciences, Beijing.

The following text is in English:
Editorial Board of China Agriculture Yearbook (1987). *China Agriculture Yearbook 1986 (English Edition)*. Agricultural Publishing House, Beijing.

8

Summary and conclusions: the characteristics and problems of Chinese agriculture and prospects for the future

XU GUOHUA

The Characteristics of Chinese agriculture

The main characteristics of Chinese agriculture are:

1. A long history of cultivation. Cultivation has been practised in China for at least six thousand years, and when it is considered that Chinese civilization has been continuous, it follows that there are numerous fields which have already been cultivated for thousands of years, and that Chinese peasants have accumulated a very rich experience of farming.

2. A large population and limited cultivated land. There are more than one billion people in China at present, and only about one hundred million hectares of cultivated land. (The figure published officially is less than one hundred million hectares, it is said that the actual area is larger than that, but no formal correction has yet been announced). That means that the per caput area of cultivated land available to the Chinese people is only about 0.1 ha. The per caput figures for the rural population and agricultural labour force are 0.12 and 0.26 ha respectively.

3. The importance of crop cultivation. In China crop production, especially 'grain' production, has always occupied a dominant position in agriculture. According to the figures published by the State Statistical Bureau, the value of crop production was nearly three times greater than that of animal husbandry, 15 times that of fishery, and 12 times that of forestry in 1986. In the same year the sown area of 'grain' crops was more than three-quarters of the total sown area.

4. The predominance of multiple cropping. The major part of the cultivated land in China is utilized under multiple cropping systems. Only in places where the altitude is too high or the frost-free period is short are the peasants prevented from growing more than one crop per year. If it is at all possible to grow more than one crop a year the peasants will do so.

5. Meticulous cultivation. Compared with cropping methods in many other countries, Chinese methods feature a larger number of different types of cultivation practices and these tend to be carried out more frequently. Consequently the Chinese peasants tend to devote more labour to a given area of land than do cultivators in other lands. In most cases, when a Chinese person says that Chinese cultivation is meticulous, he or she says it with pride. But this needs some analysis as it may be questioned whether all the practices the Chinese peasants undertake in growing their crops are really necessary. It is possible that a considerable number of them are useless or even harmful.

6. A low level of mechanization. Nearly all the agricultural practices were carried out by manpower or by animal power until the end of the 1940s. Since then the situation has changed considerably as the agricultural machinery industry has developed. It is now no longer rare to see some kind of agricultural machine operating in a rural area. However, even now, there are still many places where the agricultural practices are mainly carried out by man or animal power. Tractors, especially walking tractors, are no longer scarce in the rural areas, but they are largely used for transport rather than cultivation.

7. Wide use of organic manure. Until recently organic manures were the main sources of fertilizer for crops. The organic manures the peasants use include human and animal excrement, green manure, compost, and river-bed and pond-bed sludges. Although these manures are still used today, the amounts are decreasing, while the amounts of chemical fertilizers applied have increased rapidly over the last 30 years or so.

8. Regional variation. The regional differences are very great. The eastern part of China is completely different from the western part of China. Indeed, more than 95 per cent of the country's population and economic activity are in the east, although the land area is about the same as that of the western part. There are also obvious differences between the northern and southern parts of the eastern half of China. The southern part of east China is an agricultural region in which rice growing is the dominant activity, while the northern part of east China is an agricultural region where the main crops are wheat, maize, and cotton. Not only are the crops different, but the landscapes, living conditions, and even the life styles of the people differ.

These are the main characteristics of the agriculture of China; the most important is the large population and limited cultivated land. Many other characteristics derive from this. For example, it is obvious that under the circumstances of a large population and a small area of cultivated land per person, the first thing the peasants have to do is grow 'grain' crops for food and this is always the first priority of the peasants. It is only after this need has been satisfied that other activities can be considered. It is well known that even though the Chinese emphasize 'grain' production in this way the 'grain' production per head of the whole population was only 396 kg in 1984, the year in which 'grain' production reached its highest level ever. What would happen if the Chinese did not emphasize 'grain' production in this way?

The popularity of multiple cropping is also due to land scarcity. The peasants seek to grow more crops in a year on their limited land in order to harvest more 'grain' for food. Although the return per unit of labour is usually lower in multiple cropping systems compared with single cropping systems, the total return of crop product is generally higher.

Meticulous cultivation is a further example. Labour spent in this way on a certain crop, or on a given area of land, in general yields a diminishing return. The reason that the Chinese peasants always cultivate meticulously may be attributed to a lack of other outlets for their labour. Peasants cultivate meticulously in order to gain more agricultural products even if this involves a diminishing return for the increased labour inputs. If more arable land was available for cultivation it is probable that meticulous cultivation would be less beneficial than cultivating more land. In other words, it is essential to bear in mind the large population and limited cultivated land in order to understand the agriculture of China.

The changes in Chinese agriculture over the last thirty years

Chinese agriculture has experienced large changes over the past 30 years. The infrastructure related to irrigation, flood control, and drainage — incorporating dams, dykes, canals, ditches, and pumping stations — is much more extensive than it was in the 1950s. Greater areas of land are now irrigated, and greater areas are now protected from floods and waterlogging. Droughts, floods, and waterlogging still bring disaster every year, but the situation is far better than it was, and agricultural production has increased as a result of these improvements.

Fertilizer and pesticide production and usage have increased rapidly, in parallel with the growth of industry, and have certainly made a positive contribution to agricultural production. Disasters from vermin were common in the past, with locust plagues causing the most serious disasters of all. Since the middle of the 1950s such plagues have disappeared. In addition, the amount of plastic film used in agriculture for ground mulching, and for tunnels and greenhouses, has recently increased very rapidly, with a consequent rise in production.

Other improvements of major importance have been the development of the agricultural machinery industry which now provides many items of machinery for agricultural use; and the extension throughout the country of many good varieties and hybrids of seeds, most of which were bred in Chinese research institutions and some of which were introduced from abroad. Improvements in transportation, communications, and the power supply have also had a very positive effect on the development of the countryside.

With all these changes, and the high morale of the peasants, agricultural production has risen considerably. The Gross Output Value of Agriculture in 1952 was 46 billion yuan, and this had risen to 458 billion yuan in 1985. Even when calculated at fixed prices the increase was still 452 per cent from 1952 to 1985. These achievements are impressive, despite certain errors in agricultural and rural policy in the course of agricultural development.

But are these changes all good for Chinese agriculture and the Chinese people? Most undoubtedly are, though some may warrant further consideration. For example, the extensive use of chemical fertilizers and pesticides may bring about some harmful effects on the environment. Those who advocate ecological or organic agriculture believe that the widespread use of chemical fertilizers and pesticides is harmful and laud instead the original, traditional type of Chinese cultivation that mainly relies on organic manure and man and animal power. It is argued that the reason the land of China can be kept fertile after thousands of years of cultivation is that the people of China cultivate their land in the traditional way, which is ecologically sound. The well-known book published in 1911, *Farmers of Forty Centuries* by F. H. King, an American, gives the typical presentation. According to this opinion, the extensive application of agricultural chemicals would lead to disaster in the long term. They might be right. At least their spirit of considering things in the long term is admirable, and this is a problem that has to be investigated carefully and seriously. On the other hand, when faced with the problem of how to feed a population of more than one billion people, there seems to be no other way than to use more fertilizers.

The problems facing Chinese agriculture

The developments of the last 30 years, and particularly of the last 10 years, have brought some problems which are a cause for concern.

1. The loss of cultivated land. As a result of industrialization, urbanization, and population growth, the amount of land used for non-agricultural construction has increased rapidly; large areas of cultivated land have been taken over and lost to agriculture. During the period 1978–85, 2 666 667 ha of cultivated land were taken over for other purposes, and this represents a loss of 400 000 ha of cultivated land each year. The loss of cultivated land in 1984 and 1985 was even more serious, at an average of 733 333 ha each year in the country as a whole (*People's Daily*, 17 and 18 February 1987). In Anhui province the area of cultivated land decreased by 1 280 000 ha in the last 35 years (*Journals Digest*, 5 May 1987). In Shanxi province in 1954 there

were 4 688 000 ha of cultivated land, but this decreased to 3 760 667 ha by the end of 1985, a decrease of 926 667 ha in the 31-year period; Sichuan province lost 188 667 ha of cultivated land in 1984–85 (*Lookout*, no. 14, 1986).

2. Desertification. Desertification in the arid and semi-arid western part of China is a serious problem. The deserts increased by 40 000 km^2 over the last 25 years to the end of 1986 (*People's Daily*, 17 February, 1987).

3. Soil erosion. It was reported in the 1950s that the area suffering from water erosion was 1.5×10^6 km^2, and the area suffering from wind erosion was 1.3×10^6 km^2, throughout China; 29 per cent of the country was being damaged by erosion of one form or other. The situation has changed little in the past 30 years, and in some places has become even more serious. The annual amount of silt transported down the Yellow River has remained at about 1600×10^6 tonnes, the same as in the 1950s. But if it is considered that there have been many reservoirs built in the past 30 years, and that there has always been a great quantity of silt deposited in the reservoirs, then the unchanged amount of silt being transported in the main river implies that the soil erosion in the middle reaches of the Yellow River has actually increased. The problem of soil erosion in the Yangtze River Valley is also quite serious. In Jiangxi province in the 1950s the area suffering from soil erosion was 1.07×10^6 ha; by 1964 this had expanded to 1.80×10^6 ha, and by 1985 it had reached 3.46×10^2 ha or 20.7 per cent of the total land area of the province. In Sichuan province the area now in need of urgent soil erosion control has reached 140 000–170 000 km^2, which is 25–30 per cent of the total area of the province (*Soil and Water Conservation Bulletin*, no. 8, June 1986).

4. Environmental pollution. The primary source of pollution is industrial and domestic sewage. With the growth of industry the quantity of industrial sewage and other waste grows ever larger. Until now, most of this sewage has been discharged into rivers and lakes without any treatment, which has led to large areas of contaminated water. In addition, large amounts of industrial residues occupy a great deal of land and result in environmental pollution there. The industries that easily contaminate the environment are gradually being moved to the countryside, and in recent years the contaminated areas have been growing rapidly (*People's Daily*, 17 February, 1987). A second form of pollution is that caused by the increasing use of fertilizers and pesticides. As the application of agricultural chemicals increases so too does the extent of soil pollution and water contamination.

5. Shortage of water resources. The natural water resources of northern China are insufficient. As the irrigated area was doubled several times in the past thirty years, the amount of water used in agricultural irrigation has increased substantially. At the same time the growth of industry and of urbanization has led to increases in the amount of water consumed for industrial and domestic purposes. Therefore the shortage of water resources is becoming serious, and in some places it is very serious. There are many

places in North China where the water table has dropped rapidly due to the excessive exploitation of the ground water. There is only little water stored in many reservoirs. Many famous lakes in Hebei province have lost their water completely and the lake beds now form crop lands. In fact, all rivers north of the Yellow River, including the north section of the famous Grand Canal, have little or practically no water flow except during the flood season.

There are other problems such as the decrease in soil fertility and deforestation. Of the problems mentioned above, the loss of cultivated land is occurring in all agricultural regions. Relatively speaking it is more serious in the suburban areas, economically advanced areas, and densely populated areas; in the remote countryside it is less serious. The problem of desertification occurs in the north-west of China. Soil erosion is a common problem, with water erosion dominant in the eastern part of the country and wind erosion in the north-west. The soil erosion on the Loess Plateau has always been very serious, and the erosion in Sichuan province is becoming serious and attracting the attention of many people. Some people are even warning that the Yangtze River must not be allowed to become a second Yellow River. Environmental pollution is mainly a problem in the suburban and industrial areas, though there is a tendency for this problem gradually to spread to the countryside. The shortage of water resources is mainly a problem of northern China, especially of North China.

From the long-term view, these problems will have to be solved sooner or later if Chinese agriculture is to continue to develop.

The prospects for Chinese agriculture

If conditions are normal, and progress continues with the 'Four Modernizations', namely industrial modernization, agricultural modernization, scientific and technical modernization, and cultural and educational modernization, then the technology, capital, infrastructure, and inputs related to agriculture will surely continue to increase and improve; agricultural policy will continue to become more beneficial to peasants and managers, and it can be predicted that Chinese agriculture will develop steadily and the living conditions of the people of China will increasingly improve. But Chinese agriculture will always show its Chinese characteristics no matter how it develops. It will not become similar to the agricultures of many European or American countries, for they are not comparable.

As previously mentioned, Chinese agriculture has many characteristics of which the combination of a large population with limited cultivated land is the most important. This is unlikely to change in the foreseeable future and will continue to have a determining impact on Chinese agriculture.

This characteristic of a large population and limited land has two aspects. The first aspect, of the large population, obviously can not change within the foreseeable future as even with the implementation of the strict birth control policy of one child per couple, the Chinese population will still increase to 1.2 billion by the year 2000; indeed it will still continue to increase until the middle of next century. To decrease the population by means of emigration is unimaginable because of the huge numbers of people involved.

Turning to the second aspect, given that the western part of China is too dry or too high, and there are too many mountains and hills in the eastern part, the additional arable land which could be cultivated economically using present techniques is quite limited. According to estimates, it amounts to only 13 million ha. Even if all this land were cultivated, the increase in cultivated land would only be 14 per cent.

Given that the population is still increasing, whether the availability of cultivated land can remain at 0.1 ha per person is still in question, one does not have to say how difficult it would be to increase that figure.

The problem is not solely that the total population is too large and the area of cultivated land insufficient, it is also that, compared to the size of the rural population, the opportunities for non-agricultural employment are still very limited. If a large proportion of the people became engaged in sectors of the economy other than agriculture, then a large population and limited cultivated land would not matter, because the population making a direct living from agriculture might not be very great. It is well known that the percentage of the population engaged in agriculture in developed countries is quite small — none have percentages above 20 per cent. Under these circumstances the area of cultivated land per person may not be very large, but the area per farmer is still sufficient.

Unfortunately, in China the peasants are still by far the largest group in the total population. It is said that about 77 million people (45 million adult labourers and their families) have been transferred from agriculture to the non-agricultural sectors as a result of the development of the township enterprises. That is surely a marvellous achievement. But, when it is considered that the national population increases by 10 million people each year, it is easy to see that it is very difficult to reduce substantially the percentage of the population which lives by farming. Even if, in some way, the non-agricultural population were to reach 800 million in the year 2000, there would still be 400 million people making their living from agriculture and the area of cultivated land per person, for the agricultural population, would still only be 0.24 ha, or only twice what it is now. What is important is that the peasants must create production value through some other activities, otherwise an increase in productivity as a whole, and in living conditions, can not be realized.

There appear to be only two outlets for Chinese agriculture. One is to raise yields per unit area of land, and the other is to develop township enterprises. However, because the population base is so large, the percentage of the population engaged in agriculture can not be decreased substantially within the foreseeable future, so the dominating characteristic of a large population and limited cultivated land will still remain for Chinese agriculture. This is why Chinese agriculture cannot develop to be the kind of agriculture that only employs a very small proportion of the total population, while producing very large amounts of agricultural produce, as in Western countries.

References

The following texts are in Chinese:

Chinese Academy of Agricultural Sciences (1984). *The Regionalization of Chinese Plant Industry*. Agricultural Publishing House, Beijing.

Chinese Society of Soil Science (1983). *Proceedings of the 5th Conference of the Chinese Society of Soil Science*. Chinese Society of Soil Science, Nanjing.

Hua Shu (in press). Chinese agriculture. In the volume *Agriculture of the Chinese Encyclopaedia*. Chinese Encyclopaedia Press, Beijing.

State Statistical Bureau (1986). *Extracts of China's Statistics 1986*. Chinese Statistical Publishing House, Beijing.

The following texts are in English:

Editorial Board of China Agriculture Yearbook (1986). *China Agriculture Yearbook 1985 (English Edition)*. Agricultural Publishing House, Beijing.

Editiorial Board of China Agriculture Yearbook (1987). *China Agriculture Yearbook 1986 (English Edition)*. Agricultural Publishing House, Beijing.

King, F. H. (1911). *Farmers of Forty Centuries or Permanent Agriculture in China, Korea and Japan*. Republished by Rodale Press, Inc., Emmaus, Pennsylvania. (no date)

Appendix 1 Some geographical and agricultural terms

Geographical terms

(The regions described below only have approximate boundaries.)

Central China: a very vague term which denotes generally the middle part of the Yangtze River Valley and some areas north and south of that area.

Great Wall Line: the line of the Great Wall; the −25 °C mean minimum temperature isotherm runs approximately along this line. It separates the main cropping to the south from the pastoral activities to the north; and the multiple cropping in the south from the single cropping in the north.

Huang-Huai-Hai Plain: the North China Plain plus some of the area south of the Huai He River.

Inner Mongolia: lies on the north border of China, extending from North-East China to North-West China.

Loess Plateau: area with a loess layer on the ground surface; the thickness of the loess layer decreases from about 200 m in the west to a few metres in the east. The plateau lies in the north-western part of North-China, with an altitude of about 1000–1500 m.

Nan Ling Line: the line of the Nan Ling Mountain Range; the 0 °C mean minimum temperature isotherm runs approximately along this line. Tropical crops only grow well to the south of this line.

North China: approximately the area between the Great Wall Line and the Qin Ling Ridge-Huai He River Line.

North China Plain: the part of North China which lies east of the Taihang Shan Mountains and south of the Loess Plateau.

North-East China: formerly Manchuria, or the north-eastern three provinces, which now also includes the eastern part of Inner Mongolia.

North-West China: the area west of North China and north of the Qinghai–Tibetan Plateau.

Northern China: the area north of the Qin Ling Ridge–Huai He River Line. In most cases the term denotes only those areas in the eastern part of China, and does not include those areas in North-West China.

Qin Ling Ridge–Huai He River Line: the line of the Qin Ling Ridge and the

Huai He River. The -10 °C mean minimum temperature isotherm runs approximately along this line. Chinese scientists assume that evaporation and precipitation are equal along this line, and therefore that the aridity, K, here, is equal to 1. It divides northern China from southern China.

Qinghai–Tibetan Plateau: the largest and highest plateau in the world. It lies in the south-west part of China and occupies about a quarter of the total area of the country. It consists of the territories of Qinghai province and the Tibet Autonomous Region.

South China: the area south of the Nan Ling Line.

South-West China: the area east of the Qinghai–Tibetan Plateau and west of the Yangtze River Valley/South China. It is in fact the south-western portion of the eastern part (or the main part) of China.

Southern China: the area south of the Qin Ling Ridge–Huai He River Line in the eastern part of China. It consists of three regions — the Yangtze River Valley, South-West China, and South China.

Tibetan Plateau (or Tibet): the Tibetan Plateau is the southern or main part of the Qinghai–Tibetan Plateau; it forms the Tibetan Autonomous Region.

Yangtze River Valley, or the middle and lower reaches of the Yangtze River: the area south of the Qin Ling Ridge–Huai He River Line, north of the Nan Ling Line and east of the Yunnan–Guizhou Plateau. It is the main part of southern China.

Yunnan–Guizhou Plateau: the main part, or the southern part of South-West China. It consists of the territories of Yunnan and Guizhou provinces.

Agricultural terms

Coarse 'grain': is all other 'grain' apart from rice and wheat.

Crop yields: are expressed either as kg/ha of cultivated land/year, or as kg/ha sown.

Cultivated land: is land that has been cultivated at some stage.

Fine 'grain': in general means rice and wheat only, or more precisely husked rice and wheat flour. But in southern China it only means rice.

'Grain': is rice, wheat, maize, sorghum, millet, and other miscellaneous grains, as well as potatoes, sweet potatoes (but not taro or cassava) and beans, including soya beans (dried beans without pods). The potatoes and sweet potatoes are included in the total for 'grain' production after converting their weight on the basis of 5:1, that is 5 kg of fresh tubers are counted as equivalent to 1 kg of other 'grains'. This ratio has been used since 1964. Up to the end of 1963 the ratio used was 4:1. Output of all the other 'grains' refers to grain that has been threshed but not milled or treated further.

MAAHF: Ministry of Agriculture, Animal Husbandry, and Fishery. The name has now been shortened to just Ministry of Agriculture.

Mu: a measure of land area.

   1 mu = 0.0667 ha

   1 ha = 15.00 mu

   1 mu = 0.1647 acres

   1 acre = 6.07 mu

Also the name of a tillage implement. The area measure and the tillage implement are represented by different characters in Chinese, but both are translated into English as 'mu'.

Raw 'grain': is 'grain' that has been threshed but not milled or treated further.

Sown area: = cultivated area × cropping index. For example, if two crops of rice are grown on 1 ha of cultivated land in 1 year, then the sown area of rice is 2 ha; if another crop is grown after the two rice crops then the sown area for the plot for that year is 3 ha.

Appendix 2   Economic terms

The major economic indicators used in China differ from those used in capitalist countries, both in theory and in methods of calculation. The indicators mentioned in this book are briefly described below. These descriptions have been derived from information in the references listed at the end of this Appendix and are not intended to be definitive definitions.

The economic reforms are also frequently referred to in the book. It is customery to regard the guidelines introduced in December 1978 at the Third Plenum of the 11th Central Committee of the Chinese Communist Party as the starting point, with the new policies being brought into effect from 1979 onwards. Details will be found in Robert F. Ash (1988). The evolution of agricultural policy. *The China Quarterly*. 116, pp. 529–55.

GOVA: Gross Output Value of Agriculture
Agriculture in China comprises cropping, forestry, animal husbandry, sideline activities and fishery. And for economic purposes it includes the production of specialized state farms, agricultural experimental farms, rural economic organizations, and individual peasants.

The Gross Output Value of Agriculture is the total volume of products and by-products produced by agriculture expressed in value terms. The amount of each product or by-product is multiplied by the price to obtain the output value of each single item, and the Gross Output Value of Agriculture is the total of the values of all the individual items.

Changes were made to the items included in the calculations in 1957, 1958, 1980, and 1984. In particular, from 1984 onwards, industries run by villages, by individuals, and as joint ventures have been included in the industry sector instead of the agriculture sector.

NOVA: Net Output Value of Agriculture
The Net Output Value of Agriculture is the newly created value resulting from agricultural production. It is calculated by subtracting the value of the 'material consumption' during agricultural production from the Gross Output Value of Agriculture.

There are three major components of the material consumption of agriculture: the materials actually consumed, such as seeds, feed, manures, fertilizers, fuels, and the purchase of small farm tools; depreciation for the wear and tear of assets (which in China are called 'fixed' assets), such as farm machinery, equipment, draught animals, and buildings; and service expenditure incurred, such as machinery and equipment repairs, management, transport, postage and telecommunications, and other services received for production.

GOVI: Gross Output Value of Industry
The Gross Output Value of Industry is calculated in the same way as the GOVA, and includes the output of state-run industry.

The industry sector in China comprises the extraction of natural resources, such as mining, salt harvesting, and felling trees at above village level (but not tree planting); the processing of farm and sideline products, such as rice husking, wheat milling, wine making, cotton ginning, and animal slaughter; the manufacture of industrial products, such as steel making, spinning and weaving, machinery manufacture, dress making, and paper making; the repairing of industrial products, such as repairing machinery and transport vehicles; and the provision of water and gas supplies, and electricity generation. From 1958 onwards, only the processing charges, instead of the whole costs of the products, have been included for such items as rice husking, wheat milling, cotton ginning, and animal slaughter.

GOVRI: Gross Output Value of Rural Industry
The Gross Output Value of Rural Industry is the value of the products produced by enterprises of township industry (previously called commune industry), village industry (previously called brigade industry), and industry at below village level, which is former production team industry, industry jointly sponsored by peasants, and industry carried out by individuals. It includes the value of finished and semi-finished products, and of services provided to other enterprises.

The products covered should meet either the quality standards set by the competent authorities of the State Council or provincial governments, or the technical specifications of the order contracts. Products that fail to meet these standards should not be included. For industrial operations that restore or improve the value of existing products, or which constitute part of the whole production process, only the processing charge will be included.

TPS: Total Product of Society
Also called Global Social Product. The Total Product of Society is the sum of the Gross Output Value of the following five 'material production' sectors: agriculture, industry, construction, transport, and commerce (including catering and the supply and marketing of materials).

Material products are produced directly by the sectors of agriculture, industry, and construction, while the sectors of transport and commerce add value to the products by continuing the process of material production.

The Total Product of Society does not include the value of production from the non-material production sectors such as culture, education, public health, scientific research, public services (barbers, public baths, photo studios, hotels etc.), government agencies, police, and the armed forces.

RTPS: Rural Total Product of Society

The Rural Total Product of Society is the sum of the Gross Output Value of the five material production sectors in the rural areas. The two major components are the Gross Output Value in the five material production sectors produced by rural co-operative economic organisations at and below township level, and by individual households; and the Gross Output Value of Agriculture produced by state farms.

The Gross Output Value produced by state-owned units in industry, construction, transport, and commerce in rural areas is not included in the Rural Total Product of Society, and nor is the Gross Output Value produced in these four sectors by state farms, or by counties and towns.

NI: National Income

The National Income refers to the newly created value, created in the five material production sectors of the economy. It is the sum of the Net Output Value of the agriculture, industry, construction, transport, and commerce sectors, and can be obtained by deducting the material consumption of these sectors from the Total Product of Society.

In China two approaches are used in calculating National Income, the production approach and the distribution approach. In the production approach, National Income is obtained by subtracting the value of the material consumption (see NOVA above) from the Gross Output Value of each of the five material production sectors to give the Net Output Value for each sector. In the distribution approach, National Income is taken as the sum of the payments received by the workers in the material production sectors, plus profits, taxes, and interest in these sectors.

The non-material production sectors are excluded from the calculations of National Income.

TIRE: Total Income of the Rural Economy

The Total Income of the Rural Economy is the income received by rural economic organizations and peasants which can be used to defray expenses and is available for distribution between the state, the collectives, and individuals. It includes income from agriculture, township-run industry, construction, transport, and commerce as well as interest and rents received. It does not include income which cannot be used to defray expenses or be distributed, such as loans, advances, state investment, and investment by peasants.

The Total Income of the Rural Economy, and the Gross Output Value of

Agriculture are two different concepts. The TIRE includes disposable income from activities not included in the GOVA; the TIRE only includes that part of the income which can be spent and distributed; GOVA is calculated on the basis of unified constant or current prices of products, while the TIRE is calculated on the basis of current selling prices if the products are sold, or on the current local purchasing prices if the products are consumed by the producers.

Commodity Rate
The commodity rate is the percentage of a product that is sold on the market, rather than being consumed by the household itself.

Production
Production statistics normally include that part of the production consumed by the producer.

Township Enterprises
The township enterprises are the enterprises previously run by the rural collectives at two levels, that is, the rural people's communes and the production brigades. After the rural administrative functions were separated from the production and operation activities, in addition to enterprises run by rural cooperative organisations at township and village levels, some joint enterprises of commune members, cooperative enterprises of other forms, and individual enterprises also started to emerge. It was decided in March 1984 that all the above-mentioned enterprises should be called 'township enterprises'.

As a statistical unit a township enterprise should have a fixed identity and location, maintain production equipment and a labour force, have an accounting system, undertake economic responsibility, and commit itself to paying taxes.

Yuan
The currency of China is the Renminbi (Rmb) and the basic unit is the yuan. In 1985 2.94 yuan = US$1.00 and in 1988 3.72 yuan = US$1.00

References

The following texts are in English:
State Statistical Bureau (1986). *China: A Statistical Survey in 1986.* New World Press, and China Statistical Information and Consultancy Service Centre, Beijing.
State Statistical Bureau (1987). *Yearbook of Rural Social and Economic Statistics of*

China 1986. China Reconstructs, and China Statistical Information and Consultancy Service Centre, Beijing.

State Statistical Bureau (1988). *China Statistical Yearbook 1988*. International Centre for the Advancement of Science and Technology Ltd, Hong Kong, and China Statistical Information and Consultancy Service Centre, Beijing.

Appendix 3 Population and labour statistical terms

Extracts from the official Chinese statistics have increasingly been published in English versions over the last 4 years by the Agricultural Publishing House, Beijing, the China Statistical Information and Consultancy Service Centre, Beijing, and other official organizations. In English language publications of this type relating to agriculture, a number of words, such as grain, rural, urban, agricultural, non-agricultural, peasant, and worker, have meanings that are unfamiliar to native English speakers. However, it is beyond the scope of this book to go into these matters in detail as an explanation of them would entail an explanation of the Chinese political and economic system and how it is enumerated, and this would take many pages. A few definitions, though, are given below in the hope that they may help the reader in understanding some of the material presented in earlier pages.

Agricultural population in 1978, as referred to in Chapter 4
In China today people are designated, when registered at birth, as belonging either to the agricultural population or to the non-agricultural population, according to the designation of their mother. Originally this reflected the source of livelihood of the people, but, particularly since the introduction of the economic reforms in 1979, increasing numbers of people designated as belonging to the agricultural population have obtained their living from non-agricultural activities.

It has always been deemed that members of the registered agricultural population have access to land and therefore the means of subsistence, and that members of the non-agricultural population do not have access to land and therefore do not have the means of subsistence and so must be given a subsidy of 'grain' and other assistance by the state to enable them to live. The total number of people classified as non-agricultural, therefore, is of considerable importance to the government as it must provide subsidies for them.

It can be assumed that in 1978 most of the people counted in the registered agricultural population were living in the rural areas and obtaining their living either from agriculture or from closely related activities. The registered agricultural population of China in 1978 was recorded as 810 million people and the total population of the country as 963 million people.

During the 'cultural revolution' the collection of rural statistics was seriously disrupted and by 1978 the registered agricultural population figures still provided the most readily available estimate of the rural population.

Labour force employed in the national economy
This term refers to the whole labour force undertaking work and receiving income in cash or kind therefrom. It includes all the workers and staff of state-owned units, various joint ownership units, and collective units in cities and towns, the individual self-employed workers in cities and towns, and the peasants in townships and villages. Also included are township or village labourers undertaking household sideline activities with income equalling the lowest income level of the local labourers, and urban young people who have had temporary work for more than three months and whose income equals the wage level of the local grade-one workers.

Retired people who have been re-engaged to work, urban teachers at non-government schools, and urban labourers participating in work in the countryside are not included in workers and staff or in the rural labour force, but as they all participate in work in society and receive some payment, they are included in the labour force employed in the national economy under the item 'other'.

National Population Censuses
1 July 1953 — first census
1 July 1964 — second census
1 July 1982 — third census

Rural population and rural areas
Throughout this book, unless otherwise stated, these terms include the rural population and rural areas to be found within the designated boundaries of the cities and towns. The main exception is in Chapter 2 where the administrative divisions are discussed, and the alternative definition of rural, of outside the designated boundaries of the cities and towns, is used — see Tables 2.3 and 2.4.

Town
This term refers to a town approved by the administrative government of a province, autonomous region, or municipality. Before 1963 a town was generally defined as an area inhabited by at least 2000 permanent residents, of whom 50 per cent or more were non-agricultural residents. However, in 1964 a town was redefined as an area inhabited by at least 3000 permanent residents, of whom 70 per cent or more were non-agricultural, or an area inhabited by 2500–3000 permanent residents, of whom 85 per cent were non-agricultural.

Since 1984 a new definition has been adopted by which a town may be established as long as it is also the location of the county government; or it is a township with a population of over 20 000 where the township government is located and the non-agricultural population is over 10 per cent of the township population; or it is in an area where it is necessary to establish a town. This latter may include a place where the non-agricultural population is less than 2000 in a minority region, a far-away area with a low population density, a mountainous area, an area with a concentration of factories or mines, a small port, a scenic tourist spot, or a frontier settlement.

Workers and Staff
This term refers to all kinds of people receiving salaries or wages from enterprises, institutions, organizations and their subsidiaries of various economic forms, including those under state ownership, collective ownership, state-collective ownership, state-individual ownership, collective-individual ownership, and Chinese-foreign joint ventures, as well as enterprises or institutions run by overseas Chinese businessmen from Hong Kong and Macao, and foreign businessmen.

Workers and Staff of State-Owned Units
This refers to all kinds of personnel working for government agencies, people's organizations at various levels, and state-owned enterprises and institutions under these agencies and organizations.

References

The following text is in Chinese:
State Statistical Bureau (1988). *China Population Statistics Yearbook 1988*. China Outlook Press, Beijing.

The following texts are in English:
State Statistical Bureau (1988). *China Statistical Yearbook 1988*. International Centre for the Advancement of Science and Technology Ltd, Hong Kong, and China Statistical Information and Consultancy Service Centre, Beijing.
Yu Min (1988). China's rural socio-economic statistical system. In Heung Keun Oh (Ed.) *Development of Food & Agricultural Statistics in Asia and Pacific Region, 1965–1987*. Korea Rural Economics Institute, Seoul. D49/1988.7. pp. 277–90.

Appendix 4 Sown areas of crops, plantations, and orchards, numbers of livestock, and production of animal fibres and silkworm cocoons, by province, 1986

For a description of how the statistics which follow were obtained, see Yu Min (1988). China's rural socio-economic statistical system. In Heung Keun Oh (Ed.) *Development of Food & Agricultural Statistics in Asia and Pacific Region, 1965–1987*. Korea Rural Economics Institute, Seoul. D49/1988.7. pp. 277–90. (In English)

Appendix 4

TABLE 1.   Sown areas of crops by province, 1986 (thousand hectares)

Province	Rice				Wheat		Tubers		Maize	Sorghum	Millet
	Early	Mid-season & late	Double-crop late	Northern	Winter	Spring	Potatoes	Other tubers			
Beijing				40	183	2	1	8	216	7	12
Tianjin				35	131	13		7	147	43	14
Hebei				123	2483	9	96	351	1900	222	631
Shanxi				8	996	51	209	44	567	163	381
Inner Mongolia				27		937	225		588	133	414
Liaoning				501	1	20	24	42	1258	441	206
Jilin				351		50	88	5	1990	164	218
Heilongjiang				507		1969	209		1689	175	410
Shanghai	53	143	89		73				10		
Jiangsu	61	2278	84		2266			275	483	16	
Zhejiang	1036	182	1138		302		22	115	42		
Anhui	681	925	578		1954			658	246	61	1
Fujian	595	307	582		83			217	3	2	
Jiangxi	1559	257	1435		87			105	8	1	2
Shandong				107	4218			818	2244	123	209
Henan				412	4638			783	1885	100	195
Hubei	738	983	819		1306		172	186	282	10	10
Hunan	1838	499	1990		177		28	266	104	9	
Guangdong	1723	58	1825		41		15	579	42	1	1
Guangxi	1158	181	1180		13			237	478	2	3
Sichuan	56	3002	54		1995		425	1230	1618	101	
Guizhou	1	735			260		236	110	618	16	7
Yunnan	45	982	22		429		202	41	937	4	

Appendix 4
TABLE 1.   (Cont.)

Province	Rice				Wheat		Tubers		Maize	Sorghum	Millet
	Early	Mid-season & late	Double-crop late	Northern	Winter	Spring	Potatoes	Other tubers			
Tibet				1	22	14	1	1	2		
Shaanxi				156	1683	15	229	98	950	42	172
Gansu				4	746	740	250		233	21	67
Qinghai						201	32				
Ningxia				51	45	246	35		46	1	25
Xinjiang				69	759	459	11		436	18	1
Total	9544	10532	9796	2392	24891	4726	2510	6176	19022	1876	2979

Note: Winter wheat is wheat grown in winter. This category includes some wheats which wheat breeders from other countries would classify as intermediate types.

Source: Editorial Board of China Agriculture Yearbook (1988). *China Agriculture Yearbook 1987 (English Edition)*. Agricultural Publishing House, Beijing.

Appendix 4
TABLE 1 (Cont.)

Province	Soya beans	Other 'grain' crops	Cotton	Oil seed crops						Bast fibres		
				Peanuts	Rape	Sesame	Linseed	Sunflower	Other oil-seed crops	Jute & kenaf	Ramie	Hemp
Beijing	10	19	3	15		2		19	3	1		
Tianjin	38	29	18	12		9		65	1			
Hebei	370	642	707	340	23	105		113	6	10		4
Shanxi	199	527	94	25	20	20	116		99			4
Inner Mongolia	264	992			67	12	139	258	111			2
Liaoning	410	134	18	162		23	156	81	15			1
Jilin	489	115		7	30			181	85			3
Heilongjiang	2197	168		1				138	14			3
Shanghai	4	75	32		78							
Jiangsu	350	682	497	136	476	11			1	11	15	1
Zhejiang	64	265	81	8	281	5				31	5	
Anhui	704	243	206	141	768	191			4	79	15	8
Fujian	79	29		90	16	1			1	1	2	
Jiangxi	138	38	61	80	273	61				10	29	
Shandong	621	108	1010	850	3	22			5	24		5
Henan	973	386	620	379	260	280			2	67	6	3
Hubei	183	304	413	71	437	208			4	31	56	
Hunan	173	127	86	66	398	11		1	1	8	74	
Guangdong	126	53		415	20	9				10	4	
Guangxi	220	60	1	184	9	13			3	12	7	
Sichuan	198	711	118	168	847	10	1	3	2	50	55	2

Appendix 4
TABLE 1 (Cont.)

Province	Soya beans	Other 'grain' crops	Cotton	Oil seed crops						Bast fibres		
				Peanuts	Rape	Sesame	Linseed	Sunflower	Other oil-seed crops	Jute & kenaf	Ramie	Hemp
Guizhou	121	171	3	31	369	1		9	4	1	4	1
Yunnan	64	606	2	33	92	1		3	4		1	4
Tibet		150			9							
Shaanxi	217	415	54	40	150	10	11	14	43		1	1
Gansu	52	651	5		76		180	13	14			3
Qinghai		154			96		4		1			
Ningxia	23	187			3		61	6	24			
Xinjiang	9	49	276		113	1	81	134	7			1
Total	8296	8090	4305	3254	4914	1006	749	1038	454	346	274	46

Appendix 4
TABLE 1 (*Cont.*)

Province	Flax	Sugar cane	Sugar beet	Tobacco	Medicinal crops	Other industrial crops	Vegetables	Melons	Green fodder	Green manure
Beijing						1	58	11	10	1
Tianjin						3	44	10	4	
Hebei			8	10	9	14	285	94	76	15
Shanxi			13	2	3	13	112	44	77	7
Inner Mongolia			75	4	1	29	67	20	144	29
Liaoning			13	16	3	6	231	12	23	20
Jilin	2		43	21	12	14	161	21	5	1
Heilongjiang	79		306	50	2	60	251	77	89	40
Shanghai					1	7	71	18	5	20
Jiangsu		6	2	4	5	58	312	75	34	327
Zhejiang		22		3	14	9	199	47		458
Anhui		3		29	8	58	180	78	4	311
Fujian		69		22	3	32	174	17	19	54
Jiangxi		39		6	5	37	211	30	30	935
Shandong			4	83	3	76	348	121	10	13
Henan		6		198	11	30	334	176	8	63
Hubei		10		65	10	9	309	32	30	597
Hunan		29		74	8	12	306	33	116	1085
Guangdong		386		28	26	177	397	32	82	28
Guangxi		234		13	4	158	178	17	61	110
Sichuan		52	2	107	18	96	526	11	282	73
Guizhou		5		160	1	17	139	11	23	81
Yunnan		88		188	8	25	126	5	47	59

Appendix 4
TABLE 1 (Cont.)

Province	Flax	Sugar cane	Sugar beet	Tobacco	Medicinal crops	Other industrial crops	Vegetables	Melons	Green fodder	Green manure
Tibet							8		3	
Shaanxi			3	38	9	6	134	55	107	19
Gansu			16	4	16	28	60	19	247	29
Qinghai							5		12	2
Ningxia			10	1	1	3	14	7	28	3
Xinjiang			25			52	64	64	163	58
Total	81	949	520	1126	181	1030	5304	1137	1739	4438

Appendix 4

TABLE 2.   Areas of plantations, orchards, and tropical crops by province, 1986 (thousand hectares)

Province	Plantations			Orchards					
	Mulberry (for silk)	Oak (for silk)	Tea	Apple	Pear	Citrus	Vineyards	Banana plantations	Other fruit orchards
Beijing				11	7		2		11
Tianjin				7	2		2		11
Hebei	4	12		153	123		11		185
Shanxi	6			69	22		4		42
Inner Mongolia		22		10	4		1		8
Liaoning		491		198	60		6		76
Jilin		17		2	6		3		13
Heilongjiang		76		6			1		9
Shanghai					1	2	1		4
Jiangsu	102		14	28	14	4	3		22
Zhejiang	86		173	1	7	97	2		60
Anhui	23		116	17	14	3	3		13
Fujian			120	1	4	69	1	9	101
Jiangxi	2		60		4	42			7
Shandong	12	61	2	334	34		20		141
Henan	3	171	15	138	12	2	13		74
Hubei	10	21	67	7	12	53	1		7
Hunan	4		109		7	98			10
Guangdong	18		47		7	96		43	294
Guangxi	3		21		2	45		9	30

Appendix 4
TABLE 2.   (Cont.)

Province	Plantations			Orchards			Vineyards	Banana plantations	Other fruit orchards
	Mulberry (for silk)	Oak (for silk)	Tea	Apple	Pear	Citrus			
Sichuan	47	2	105	16	11	136			12
Guizhou	2	2	28	1	2	9		1	2
Yunnan	5		121	8	11	5	1	11	17
Tibet				1					
Shaanxi	33	2	25	93	7	11	6		49
Gansu	1	2	1	46	13	2	1		8
Qinghai				3	1				1
Ningxia				9	1		1		2
Xinjiang	2			17	9		35		38
Total	363	879	1024	1176	397	674	118	73	1247

Source: Editorial Board of China Agriculture Yearbook (1988). *China Agriculture Yearbook 1987 (English Edition)*. Agricultural Publishing House, Beijing.

Appendix 4
TABLE 2 (*Cont.*)

Province	Rubber	Coffee	Coconuts	Oil palm	Cashew	Pepper	Perfume crops	Sisal hemp & American aloe
Beijing								
Tianjin								
Hebei								
Shanxi								
Inner Mongolia								
Liaoning								
Jilin								
Heilongjiang								
Shanghai								
Jiangsu								
Zhejiang								
Anhui								
Fujian	8							3
Jiangxi								
Shandong								
Henan								
Hubei								
Hunan								
Guangdong	419	4	19	3	9	11	8	7
Guangxi	13							4
Sichuan								

Appendix 4
TABLE 2 (*Cont.*)

Province	Rubber	Coffee	Coconuts	Oil palm	Cashew	Pepper	Perfume crops	Sisal hemp & American aloe
Guizhou								
Yunnan	110	1					1	
Tibet								
Shaanxi								
Gansu								
Qinghai								
Ningxia								
Xinjiang								
Total	550	5	19	3	9	11	9	14

Appendix 4
TABLE 3.   Numbers of livestock by province, 1986 (thousand head)

Province	Cattle		Buffaloes	Horses	Donkeys	Mules	Camels	Sheep		
	Unimproved & draught	Improved for milking						Improved fine-wool	Improved semi-fine	Other
Beijing	55	45		39	49	76				157
Tianjin	52	16		38	86	71			3	220
Hebei	1666	59		692	1540	795		1369	423	2149
Shanxi	1469	62		159	493	518		229	139	2221
Inner Mongolia	3800	256		1824	884	475	274	6521	3860	7942
Liaoning	1360	42		561	753	423		775	582	403
Jilin	1566	36		868	146	233		1060	192	110
Heilongjiang	1567	319		1133	65	56		756	1023	74
Shanghai		52	16							52
Jiangsu	283	34	378	27	131	22				486
Zhejiang	439	48	271					485		619
Anhui	3095	11	1270	139	279	67		15	146	39
Fujian	736	26	448	1		1				
Jiangxi	1574	12	1093							
Shandong	2892	18	15	366	1128	382		1060	664	1058
Henan	6651	21	305	510	1425	662		858		314
Hubei	1722	18	1592	24	32	3		14	2	13
Hunan	1954	7	1686	8	5	2				3

Appendix 4
TABLE 3. (Cont.)

Province	Cattle		Buffaloes	Horses	Donkeys	Mules	Camels	Sheep		
	Unimproved & draught	Improved for milking						Improved fine-wool	Improved semi-fine	Other
Guangdong	2120	23	3236	220		6				2991
Guangxi	2908	5	3042							
Sichuan	6769	33	2800	440	43	23		252	149	175
Guizhou	3382	14	1705	611		8		34	137	
Yunnan	4963	39	2563	932	220	329		200	400	1060
Tibet	4940	198		296	113	11		104	297	10820
Shaanxi	1959	33	17	71	430	190		505	120	594
Gansu	2989	79		495	1329	455	42	1199	802	5922
Qinghai	5408	37		466	165	115	22	83	2430	10883
Ningxia	264	9		45	295	167	7	41	60	2519
Xinjiang	2801	294		1023	1078	23	159	10178		11015
Total	69384	1846	20437	10988	10689	5113	504	25741	11429	61839

Source: Editorial Board of China Agriculture Yearbook (1988). *China Agriculture Yearbook 1987 (English Edition)*. Agricultural Publishing House, Beijing.

Appendix 4
TABLE 3 (Cont.)

Province	Goats		Pigs	Chickens, ducks & geese	Rabbits	Bee hives
	Milk type	Other				
Beijing	14	360	1456	22330	177	41
Tianjin	13	304	768	15962	302	6
Hebei	305	3913	14074	97198	8857	176
Shanxi	124	1705	3665	35297	2639	129
Inner Mongolia	29	6670	4810	25686	603	47
Liaoning	73	322	10317	64057	1087	76
Jilin	28	39	4812	43161	228	76
Heilongjiang	106	145	5648	50820	200	133
Shanghai	1	177	2093	18027	5755	30
Jiangsu	14	4520	19257	130810	6844	225
Zhejiang	9	609	14033	62139	7284	943
Anhui	3	2386	12235	123791	6618	204
Fujian	15	589	8513	53487	4416	176
Jiangxi	3	96	13441	75042	507	229
Shandong	581	6490	16689	151207	17758	136
Henan	288	6317	15394	156414	7768	313
Hubei	2	1334	19897	142928	296	410
Hunan	5	572	25969	140203	984	256
Guangdong		411	22627	136470	983	277
Guangxi	1	635	15638	100760	147	129

Appendix 4
TABLE 3 (Cont.)

Province	Goats		Pigs	Chickens, ducks & geese	Rabbits	Bee hives
	Milk type	Other				
Sichuan	28	5165	60635	176320	19579	1039
Guizhou		1128	12140	32310	296	165
Yunnan	24	5167	17126	32938	1040	846
Tibet		5652	144	588	4	1
Shaanxi	647	2096	7792	33910	1315	256
Gansu	11	1827	5747	22926	2621	265
Qinghai		1549	880	2429	93	1
Ningxia	6	816	658	4773	92	30
Xinjiang		3896	733	13614	254	31
Total	2330	64890	337191	1965597	98747	6646

Appendix 4

TABLE 4   Production of animal fibres and silkworm cocoons by province, 1986 (tonnes)

Province	Sheep wool			Goat wool	Cashmere	Rabbit wool	Silkworm cocoons	
	Fine	Semi-fine	Other				Mulberry	Tusser
Beijing	3		157	150	5	1	41	
Tianjin		22	162	31	2	5	2	
Hebei	4721	1083	2427	1022	335	10	290	
Shanxi	806	340	1954	504	176	9	2575	461
Inner Mongolia	23312	13080	12007	1540	1359	3	2	
Liaoning	2950	1980	867	174	71	6	30	24760
Jilin	3934	676	365	9	3	1		450
Heilongjiang	2830	3593	119	30	11		1	1197
Shanghai			87	17		518	266	
Jiangsu			1552	171	2	432	79966	
Zhejiang	2497			142		498	88945	
Anhui	75	844	69	55		315	9765	
Fujian						35	124	
Jiangxi							674	
Shandong	5183	2321	2720	1908	130	637	11162	1642
Henan	3838		625	807	36	173	360	3907
Hubei	51	8	29	22		8	3897	256
Hunan				19		35	2589	21
Guangdong				18		41	20871	
Guangxi						11	3760	

Appendix 4
TABLE 4 (Cont.)

Province	Sheep wool			Goat wool	Cashmere	Rabbit wool	Silkworm cocoons	
	Fine	Semi-fine	Other				Mulberry	Tusser
Sichuan	116	214	2074	220	4	276	102664	
Guizhou	77	255	178	11		10	402	4
Yunnan	383	717	524	54		50	1600	
Tibet	101	250	7990	924	281	13		
Shaanxi	1700	205	1021	578	252	2	5270	
Gansu	3317	1687	7068	910	170	54	50	
Qinghai	228	4551	9591	355	115			
Ningxia	134	222	3430	239	122		908	
Xinjiang	33502		8375	1591	397	4		
Total	89758	32048	63391	11501	3471	3147	336214	32698

Source: Editorial Board of China Agriculture Yearbook (1988). *China Agriculture Yearbook 1987 (English Edition)*. Agricultural Publishing House, Beijing.

Index